The Knowledge Economy in India

The Knowledge Economy in India

Edited by

Frank-Jürgen Richter

and

Parthasarathi Banerjee

Foreword by Narayana N.R. Murthy

palgrave
macmillan

First published 2003 by
PALGRAVE MACMILLAN
Houndmills, Basingstoke, Hampshire RG21 6XS and
175 Fifth Avenue, New York, N.Y. 10010
Companies and representatives throughout the world

PALGRAVE MACMILLAN is the global academic imprint of the Palgrave
Macmillan division of St. Martin's Press, LLC and of Palgrave Macmillan Ltd.
Macmillan® is a registered trademark in the United States, United Kingdom
and other countries. Palgrave is a registered trademark in the European
Union and other countries.

ISBN 1–4039–0110–4

This book is printed on paper suitable for recycling and made from fully
managed and sustained forest sources.

A catalogue record for this book is available from the British Library.

Library of Congress Cataloging-in-Publication Data
The knowledge economy in India / edited by Frank-Jürgen Richter and
Parthasarathi Banerjee; with a foreword by Narayana N.R. Murthy.
p. cm.
Includes bibliographical references and index.
ISBN 1–4039–0110–4
1. Technological innovations—Economic aspects—India.
2. India—Economic conditions. I. Richter, Frank-Jürgen.
II. Banerjee, Parthasarathi.
HC440.T4 K56 2002
338.954'06—dc21 2002025816

10 9 8 7 6 5 4 3 2 1
12 11 10 09 08 07 06 05 04 03

Printed and bound in Great Britain by
Antony Rowe Ltd, Chippenham and Eastbourne

Contents

Foreword

This book is a must read for anyone interested in the knowledge economy in India; it analyses the influence of globalization and liberalization, and the impact of technology on India's economy.

Today, globalization is an inescapable fact everywhere. Globalization is about sourcing capital from where it is cheapest, producing where it is most cost-effective, and selling where it is most profitable, all without being constrained by national boundaries. Thus, every country has to bring something unique to the global bazaar. In fact, every country must build on its strengths. In this regard, there is no doubt at all that technology has a key role to play in ensuring development of the Indian economy.

The various essays in this book provide a holistic perspective of the status of the Indian high-tech sector as well as its prospects. They also provide an in-depth analysis of critical issues such as entrepreneurship, infrastructure, human resources and venture capital that form the key to any knowledge-based industry. This is all the more relevant as the knowledge infrastructure is crucial for success in today's world. Moreover, the essence of democracy is equal opportunities for all. Thus, in order for democracy to flourish, the knowledge infrastructure needs to be available to all citizens.

Overall, I find this book extremely informative and thought-provoking.

NARAYANA N.R. MURTHY
Chairman of the Board and Chief Mentor
Infosys Technologies Limited
Bangalore, India

Notes on the Contributors

Raj Aggarwal is Professor and Firestone Chair of Finance at Kent State University, Ohio. A highly regarded academic, business consultant, and speaker, he has accumulated over a quarter century of experience in academic and business administration. His current interests include international business strategy, managerial and international finance, Japanese and Asian business and finance, developing capital markets and business use of technology. He has held many elected and appointed leadership positions.

Neslihan Aydogan was appointed Visiting Assistant Professor at the Kelley School of Business, Indiana University, followed by an appointment as Adjunct Professor in the School of Public Policy and Environmental Affairs. Her interests are on issues that are related to R&D and firm organization.

Parthasarathi Banerjee holds the position of scientist at the National Institute for Science, Technology and Development Studies, New Delhi. He serves as Visiting Professor on Strategic Management at some Indian management institutes. He has published nine books, as both author and editor, and has several published papers.

Dipak Basu is Professor in International Economics at Nagasaki University, Nagasaki. Professor Basu has been a Lecturer at Oxford University and Senior Economist to the Ministry of Foreign Affairs, Saudi Arabia. He has published six books and fifty scientific papers in leading international journals and read papers at major international conferences.

S.K. Chakraborty is Professor at the Management Centre for Human Values, Indian Institute of Management, Calcutta. Professor Chakraborty combines more than three and a half decades of postgraduate teaching experience in India and abroad with four years of work experience in Indian industry. He has so far published twenty-five books, two of them being award-winning. He was conferred the Best Management Teacher award for 1994 by the Association of Indian Management Schools.

Soumitra Dutta is the Roland Berger Professor of e-Business and IT, and Dean for Executive Education, INSEAD, Frame. His current research is on business innovation and technology. He has published several books as well as over fifty articles in leading international journals.

Seema Gahlaut is the Associate Director, South Asia Programme, at the Center for International Trade and Security, University of Georgia. She is in charge of the India and Pakistan sections of the Center's multi-year Global Review (of export control systems) project, funded by the Carnegie Corporation of New York. She is the co-editor of, and contributor to, *Engaging India: US Strategic Relations with the World's Largest Democracy* (1999).

C. Gopinath is a Professor at the Suffolk University, Boston.

M.V. Lakshmi is Assistant Professor of International Relations at the GITAM Institute of Foreign Trade. She has conducted research, co-authored two books and authored and co-authored several articles on themes of social development, gender issues and international trade environment. She edits *Global Vistas*, a monthly bulletin on international trade developments published by the Institute.

Deval Parikh is a Silicon Valley based consultant with Pittiglio, Rabin, Todd & McGrath (PRTM), a leading management consultancy to technology-based business.

Dwijen Rangnekar is a Senior Research Fellow at the School of Public Policy, University College, London, with a specialization in the area of evolutionary economics of science and technology, with a specific focus on pharmaceuticals and agriculture. He has an active interest in issues concerning intellectual property rights and indigenous peoples' rights.

Bhanoji Rao is the Honorary Research Professor at the Gitam Institute of Foreign Trade, Rushikonda, Vishakhapatnam. His employment background includes Associate Professor, Senior Lecturer and Lecturer in Economics, National University of Singapore and University of Singapore, and Economist, World Bank, at Washington and Jakarta. He has been author, co-author and editor of over a dozen books, contributed chapters to many more, and over fifty articles in international journals.

Mohanbir Sawhney is the McCormick Tribune Professor of Electronic Commerce and Technology at the Kellogg Graduate School of Management, Northwestern University, and the head of the Technology and e-Commerce group there.

Jati K. Sengupta is Professor of Economics and Operations Research at the University of California, Santa Barbara. Formerly he was the Director of the Indian Institute of Management, Calcutta. He has published more than 300 research publications, most recently the book *New Growth Theory: An Applied Perspective* (1999).

Nirvikar Singh is Professor of Economics at the University of California, Santa Cruz. He is currently Director of the Business Management Economics Programme. He teaches business strategy, technology and innovation, and electronic commerce, as well as graduate microeconomic theory. Professor Singh's current research topics include governance and economic reform in India, international technology transfer, international water disputes and economic growth in East Asia.

Dipendra Sinha is with the Department of Economics, Ritsumeikan Asia Pacific University, Japan, and Macquarie University, Australia. His research fields include applied microeconomics, economic development, industrial organization and regulation, as well as international economics and international business.

Anupam Srivastava is the Director of the South Asia Programme at the Center for International Trade and Security, University of Georgia (UGA). He also serves as the co-Director of a major UGA-wide 'India Initiative' that seeks expanded cooperation between India and the USA. He also serves as a Senior Instructor for the Department of Political Science at UGA. Dr Srivastava has published widely in books, policy journals and newspapers.

Navendu Vasavada is with Lumin Asset Management, Pittsburgh. In his early career, he worked on assignments with the Indian government and with Novartis (India), as Assistant Professor in Finance at the Pennsylvania State University and as a project manager at a research centre at the University of Pennsylvania.

C.S. Venkata Ratnam is a Professor in the International Management Institute, New Delhi.

Introduction

> Every country or firm must have education and training in technology and science, even if the research is not on par with that being conducted elsewhere. Knowledge cannot be absorbed unless some knowledge is already possessed.
> Countries and firms must be open to new ideas, have multiple sources of new ideas, and see that ideas are diffused. This point strongly argues for freedom of entry, even when it seems to forgo economies of scale.
>
> (Arrow, 2000, p. 19)

The knowledge economy in India seems to be approaching norms laid out by such Arrowian imperatives. Liberalization and globalization of the Indian economy is taking place when its economy is toddling on the two legs of the old and the new – the former characterized by lack of competition and lack of openness, and the latter exhibiting germinal openness, competitiveness and knowledge intensity. Undeniably in its transition to a knowledge economy, this new economy has received its support mostly from the markets of the OECD countries while its own home market has remained saddled with the old economy struggling for its existence. Diffusion of ICT (information and communication technology) in the home market has remained pitiably low. R&D intensity of manufacturing, often used as a surrogate for knowledge intensity of an economy, also puts India to shame when one compares statistics even with Chinese counterparts. Interfacing ICT usage with indigenous languages and with indigenous cultures has thus received very low attention. This has been compounded by one of the lowest levels of literacy, threatened further by cuts in public expenditure on higher education which has failed to achieve laurels internationally over the last couple of decades. The old economy syndrome had put nails in the coffin of local entrepreneurship. Perhaps it was demands from the dynamics of the international market that Indian software could get sustenance from its fledgling software entrepreneurs. Knowledge intensity of the overall Indian economy is thus one of the lowest. Barriers and resistance to ICT diffusion have further compounded this ignorance-trap, yet India enjoys a billion head and growing population, many of whom are young. This demographic tilt is a valuable resource and perhaps the most significant political force bringing the old economy to negotiated reconciliation. The youth in their drive for knowledge-based entrepreneurial success might forget the other half of the population left behind, who may not be able to adapt to the new culture. Restraints need to be provided both culturally and institutionally to achieve a harmonious, equitable and culturally satisfying development of the Indian knowledge economy.

This volume of essays brings out this panoply and the variety of the contributions is a reflection that the Indian knowledge economy does not need a simple panegyric. The potential of ICT to become the growth-driver is clear, and the most powerful engines of this drive are the new knowledge firms. This volume begins with an examination of these aspects, and knowledge firm strategies, structures and experiences offer us a rich picture of the terrain. However, institutional constraints on diffusion and the culture of entrepreneurship keep a powerful check on this growth driver. We therefore considered that institutional aspects of entrepreneurship and the social and cultural harmony of knowledge entrepreneurship should receive our attention. The book discusses these issues then looks at certain domestic and global institutional barriers. Three barriers are discussed. The existing old economy, global high-technology barriers to trade and the domestic labour situation are considered, but all this must be reconciled with the deep roots of traditions. Globalization in the absence of values, such as once provided by post-renaissance Europe, would appear as transnationalizing and uprooting in nature. The last few essays in this volume offer some skepticism on the slogan of globalization, and faith in a possible universalization that has not forgotten its local moorings.

Background

The Indian knowledge economy has definitely identified knowledge firms as the primary driver. However, it is unsure whether this economy can develop a knowledge society as a powerful and robust driver. Knowledge firms can be sustained only if resources of skill, knowledge and entrepreneurship flow out of from society, and if flows dried up the existing knowledge firms could not be sustained. The picture of country competency and competitiveness for India is not very reassuring. A recent survey of school education in the capital city, New Delhi, sadly observed (Rajan, 2001) that only a small fraction of the age cohort attends schooling (only about 2.65 million out of a population of above 10 million) of whom 52 per cent are at the primary level, then drastically reduced to 27 per cent at the middle level, reduced further to 13 per cent at the secondary stage and finally to only 7 per cent at the senior secondary level. In parallel to this, government schools which cater to the deprived sections have the lowest percentage of students passing out, and private schools with high school fees (nearly one-third of the total of schools) have an increasing proportion of students. 'Out of every 100 children who enroll into the government and local body schools in Delhi, only 14 make it to the class X level; out of which only 4 pass the class X level' (Rajan, 2001, p. 5). The picture is worse for the country aggregate. Moreover, local communities in the countryside, especially those from tribal population groups, face a curriculum that is at variance with local/ indigenous knowledge and pedagogic tradition (Sarangapani, 2001). If this situation continues the knowledge future of India appears bleak.

Higher education, lifelong education and training, and country R&D also offers a dismal picture. Doctorates in engineering and technology are the lowest in number (ironically, boards of Indian firms have perhaps the least number of persons from technical backgrounds). Young students with better results are streaming into non-science courses. Old economy firms often do not provide any formal training or scope for life-long education. Skill mobility across vocations, data on which is not available, and personnel mobility across firms, excepting the software sector, are nearly nonexistent. This is in sharp contrast to the average OECD picture (OECD, 2001). While persons of Indian origin receiving doctorates in the USA are next only in number to those from China, when normalized over total country population and compared to a small country such as Korea this apparently satisfying figure drops down to nearly the base level. Moreover, private sector investment in R&D is nearly zero (this is sadly true for the software sector as well) reflected in its lowest international patent holding. Strategic technology partnering, cross-border technology alliances of strategic nature, international joint research and joint patenting or even joint technology development are also nearly absent. Less than 1 per cent of GDP gets invested into R&D (nearly entire amount by the government) further burdened by the fact that regional distribution of this expenditure is highly skewed. Moreover, state government (as compared to the central) participation in research and higher education, especially when one compares statistics with Korea (or even China) again, puts the country record to shame (Chung, 2001). Finally, geographic distribution of higher education and degree colleges is slanted towards certain preferred locations only. Such a picture of the knowledge society does not even ensure that India in the near future will be prepared to adopt and adapt knowledge flows into its social and institutional fabric.

The preparedness of a society to quickly pass over its transition into a knowledge-led economy from the old economy is an important indicator, particularly for those developing and transition economies who do not possess a good infrastructure. Discussions on infrastructure ordinarily refer to the communication backbone, available bandwidth, etc. A few other indicators, ordinarily used by the OECD, such as 'secure servers' per 100 000 population, would be remarkably staggering – USA has about 27 such servers, Germany little above 5 while India, on whom there is no reliable data available, is very close to zero. Internet hosts per 100 inhabitants in USA for October 2000 was 234.2, and for India this appears to be close to zero. The price of access and the bandwidth available for internet, for example, while it is US$ 20 per month for a 10 Mbit/sec access in Sweden and when this price is normalized by purchasing power parity and compared to Indian figures – the Indian figure would appear to be astronomically high. A country strategy, such as the Japanese and German strategy regarding ISDN or the general OECD strategy regarding WLL, appears amiss in India. Digitalization of both access lines and mobile subscriptions for the entire OECD are close to 100 per cent, similar figures for India would be far off that mark. Hours spent on line (average per month per subscriber) for the 'AOL' was about 32 by beginning of 2000. Not strangely perhaps the figures for internet subscribers per 100 inhabitants is about 23 for Korea (the highest in the OECD along with Sweden), while Indian figures are ridiculously insignificant. A typical internet user in India is predominantly male, between ages 25 and 34, a graduate or a post-graduate, a business man or an executive, has monthly income above Rs. 10 000/- and at least 35 per cent of such users have credit cards (INFAC) – such a population is an insignificantly small percentage of even Indian urban population.

Apart from the fact that public investment in the telecommunications network in India is insignificant when compared to not only the major OECD countries but also with China, there is also the problem of very low value of ICT usage in Indian industry. The IT intensity index in the

USA, defined as the industry's percentage share of IT expenditures relative to industry's share of GDP (thus an index of 1 reflects no over or under spending), has the highest value of 3.5 for the financial markets, close to 3 for banking, above 2 for discrete manufacturing and less than 1 for process manufacturing, wholesale, government, etc. reflects a structural pattern and such a pattern is obviously absent in India. The data for ICT expenditure (again on which Indian country level data is not available but can be approximated) was above 8 per cent for New Zealand, Sweden, Australia, United States; however, Indian figure is incomparably small. The compound annual growth rate of ICT markets during 1992–97 was about 14 per cent for India, while for China it was close to 30, for Brazil around 33, Singapore around 16, Indonesia 15.4, etc; moreover, the absolute size of this ICT market in China in 1997 was at least 5 times that of India, and in Brazil it was again about 5 times. The Indian market was slightly smaller than even Singapore, a city-state! Interestingly, employment trend too reflects this tendency. With 1980 as the base year, employment in 14 OECD countries in ICT manufacturing and in high-technology industries reached a peak around 1986 with a value of about 115 to level off to the base year share of 100 again by 1996 – during this same period this index fell from 100 in 1980 to less than 90 in 1996. Indian employment pattern is heavily tilted in favour of services, little of which involves ICT or other hightech. Corresponding to this employment trend there has been high compound annual growth rate in R&D expenditures by several ICT firms from the OECD countries. Often R&D expenditure as percentage of sales for firms would be above 10, such as for AMD, National Semiconductor, Microsoft, ST Microelectronics, Lucent, Silicon Graphics, etc. Few Indian ICT firms spend anything on R&D.

In a comparison between the USA and the European market of ICT and its contribution to growth (Roeger, 2001), it was observed that rate of technical progress – captured primarily in the production of high-tech goods, the USA enjoyed a clear comparative advantage. Although there is no clear consensus regarding ICT's contribution to growth (Chang, 2001), studies (Bassanini, Scarpetta & Visco, 2000) on improvements in labour productivity allowing for human capital accumulation, on multifactor productivity (MFP) allowing for changes in composition of fixed capital, and on changes over embodied (especially in ICT) as well as disembodied technical progress have shown that diffusion of ICT and MFP growth rates, especially in the USA, offer positive insights for policy framing; in fact, entrepreneurship appeared to be one of the key parameters. Innovation in general, ICT use – human capital – startups and the receding share of incumbents – share of emerging areas, etc., were considered as the keys to new economy based on knowledge (OECD). In all this while in absolute terms India stands nowhere, its preparedness and the fact of its departure from the old economy trappings are, however, significant. Incumbents still retain a very strong hold over the process of structural changes, human capital, excepting such sectors as the software and biotech, would not still be considered as decidedly influencing the rate of technical (disembodied) progress, venture formations are far below the mark.... yet, things are moving forward. A critical question would definitely be how far off the critical threshold are these figures!

What is on offer

Several studies conducted on these developed OECD countries indicated the importance of labour market values exhibited by globalization, in particular the popular threat perception and the perceived loss of autonomous values, technological trade regimes, corporate restructuring, share of incumbents in the industrial sectors particularly in the high-tech manufacturing, and the overall state of competition (even including a country strategy regarding that) – and as a result, suggested policy framework conditions as well as inter-country negotiation framework conditions have now begun including these parameters. It is suggestive, therefore, that this volume brings together these few facets on the emergent knowledge economy in India. With these parameters in view contributions here have been grouped under four sections.

We begin with possibilities of growth of the high tech and in consequence the country economy with ICT as the growth-driver. However, the strength of a typical Indian firm from this high-tech area in comparison to dominant incumbent global firms is insignificant. This is further compounded by the weaknesses in the domestic institutions including the interface between the public R&D and the manufacturing, etc. The ICT sector, characterized by hypercompetition, perhaps then can be considered as an industrial sector for which a country strategy could be framed given the fact that Chinese and other East Asian ICT sectors gained immensely from such public policy frameworks. One such much-debated policy issue relates to locating the software services. An obvious choice for most domestic frameworks has been offshore development of software services, and contributors here look at these issues of country-strategy, offshore development, e-services. However, such a strategy is related to the structure of software industry. This industry structure issue too then has been examined. ICT naturally has enjoyed the overwhelming attention in this volume notwithstanding Indian capabilities in biotech and pharmaceuticals. The latter is a large area and lest we increase the size of the book beyond a reasonable limit we decided to drop that altogether and bring in as an example a topical issue of seed industry privatization.

One of the key issues relating to industry structure and state of competition is the relative dominance of the 'old', the incumbent. India demographically and in terms of the composition of its human capital as well as in terms of its entrepreneurial history should have been showered with new entrepreneurs – the fountainhead of technical innovation. The following two sections present certain selected aspects of this. Entrepreneurship in contemporary economies depends to a large extent on the states of the institutions. We therefore focus on entrepreneur development under the constraints of given status of venture funding, etc. In the following we take up three perspectives: first on how the incumbent corporate are likely to de-merge or restructure; second on how some international technology trade control regimes might affect transfer of technologies; and third on how the prevailing labour institutions might react to the threats from globalization.

A deep-rooted fear of globalization appears now in the form of often violent demonstrations against liberalization and globalization in the OECD countries. Perhaps perceptions of threats in countries such as India, China and Brazil would take up other alternate forms and expressions. Some ideologues have begun talking about social management of such threat perceptions. Alongside the threat perceptions youth in all these countries, while sharing insecurities, are also making fantastic claims. A complex interweaving of local versus global, community versus individual or the global, security versus uncertain prospects, and universal versus transnational antinomians, naturally raises questions as to what Indian tradition might offer and what are the roots of such threats, and how India could develop and grow on certain core values. The final three chapters of this volume report on this. A knowledge economy requires for its growth another 'hundred years of peace', and knowledge, unlike the skills of the post-industrial revolutionary phase based much on the body, demands attention from the mind and the intellect – both so deeply influenced by calm non-serendipitous peace of mind! We conclude this volume with this pointer to the harmony between knowledge business and our values.

High-tech sector status and prospects

Neslihan Aydogan in Chapter 1 ('India's high-tech sector growth: creating a knowledge economy') argues that nations that prosper in the new world order are those that successfully manage knowledge assets. In this context, successful management decisions that are made at the level of each firm are as important as policies made and decisions taken at national and local technological levels. The countries that have recently joined, or will join, the technology race such as India are likely to have much to learn from front-runners such as the USA and Japan. Evidently, different countries follow different trajectories for boosting the nationwide rates of innovation and productivity. However, one thing, which is more apparent than any other ingredient to this end and is common to all high-tech nations, is the ability to prosper entrepreneurial incentives. Despite the development in information technologies, one consistently observes that knowledge tends to get localized across geographical boundaries and that it does not spillover. Therefore, the question is how can we transform an economy so that knowledge assets such as entrepreneurial capital and skilled employees become intensely concentrated and feed themselves over time. It is this question that Aydogan tackles for the development of the New Indian economy.

Nirvikar Singh in Chapter 2 ('Information technology as an engine of broad-based growth in India') surveys some of the developments in India's IT sector in order to examine the prospects for broad-based growth led by this sector. Singh examines the IT sector, discussing the role of software versus hardware, the growth patterns of the software industry and software exports, and the potential problems in IT labour supply to support future growth. Singh focus on a current bottleneck for the IT sector, namely the telecommunications infrastructure. Issues considered include the basic driver of technological convergence across voice and data communications, problems with current infrastructure, innovations that have the potential to dramatically alter the economics of access to telecoms, and the evolving structure of the telecoms industry. He also examines the policy environment more closely, arguing that government policy is better focused on removing labour market distortions and infrastructure constraints, rather than providing output or export subsidies to the software industry. He closely examines the appropriateness of specific policy goals such as universal access, as well as issues of implementation of more general objectives of broader telecoms access. Finally, Singh argues for the possibilities of a broad-based IT-led growth, including increasing value-added, using better telecom links to capture more benefits domestically through offshore development for industrial country firms, greater spillovers to the local economy, broadening the IT industry with production of telecom access devices, improving the functioning of the economy through a more extensive and denser communications network, and improving governance.

Jati K. Sengupta in Chapter 3 ('Facing hypercompetition in world software markets: global strategies for India') asks how to face hypercompetition in world software markets. Sengupta identifies this as the greatest managerial challenge facing India. He points out that India has all the potential to succeed in this market, for example, technical know-how, experience and networking capabilities. It needs to develop successful global strategies consistent with the trends of technology-intensive computer industry in the Silicon Valley. His proposed strategies are broad based for the whole economy and are not specifically oriented to corporate policies. Assuming two preconditions: outward orientation of India's international trade and the existence of a common goal of steady and sustained industrial growth for both the private and public sector, Sengupta points out the three sources of growth which spring from the level of a firm, namely, competitive advantage, comparative advantage and core competence. Strategic alliances formation *inter alia* also appear important. Sengupta looks at the Indian experience and highlights global strategies as well as managerial challenges.

Mohanbir Sawhney and Deval Parikh in Chapter 4 ('Powered by India') argue for an export-led growth of e-commerce and technology firms. They point out the prospects of the e-services frontier, defined as the outsourced activities made possible by the internet. Analysing possible prospects through the two dimensionalities of 'human capital intensity' of the activity and the 'degree of digitizability' of the service, Sawhney and Parikh look at that prospect where values of both these are high. They forecast a future winning role for India in e-services. Several examples are offered. They lay particular emphasis on customization or 'getting personal'. Implementation of this strategy would require several managerial skills, of which establishing business development teams, managing customer relationships, tools and training, quality assurance and institutionalization of knowledge appear important.

In Chapter 5 Soumitra Dutta examines the strategy of 'Using offshore development to drive value: the experience of Motorola in India'. Dutta begins his argument from an examination of the Indian software industry. Software, he argues, is playing a major role in helping India to secure its position in the global economy. Offshore software development has become a necessity for firms in the West who are facing increased dependence upon software and a lack of trained software professionals. However, most corporations regard offshore software development as a risky venture and treat it primarily as a cost-reduction action. Dutta suggests that there is a better way to leverage the software excellence in India (and other offshore software development sites). The desired strategy is to use the offshore software and technology skills to drive value by stimulating product design and business innovation. Doing this is not easy and needs a unique combination of ambitious vision, determined leadership, talented people and rigorous quality processes. The experience of Motorola in successfully using offshore software development to drive global business value is detailed to illustrate the challenges in the process.

Parthasarathi Banerjee in Chapter 6 ('Coordination, strategy and capability in Indian software') looks at the flow of human resources across Indian software firms, and noting that this is an unique phenomenon evidenced at a time when Indian software experienced secular above average growth, Banerjee points out that received understanding on resources-based and capability-based perspectives of strategy formulations fail to explain firm growth with continuous flow of human resources. Site specificity of knowledge would have required a firm to stop, forcibly if necessary, such passage of resources. Moreover, human resources is the principal resource of such a firm. Banerjee draws sustenance from the received literature: human resources in software act as intermediate goods, passage of which across software firms provide the necessary information required in an industry afflicted with hypercompetition. Such information is required for three purposes, namely, as technical information, as market information and as that information required to sustain innovation. The first two satisfy demands of coordination while the last satisfies the demands of future innovation. As a result, Banerjee points out the policy relevance, by arguing that flow of human resources across firms is vital to the innovative capability, and that vertical integration or even integration along the dimension of scope would prove to be detrimental to innovative capabilities of software firms.

Dwijen Rangnekar in Chapter 7 ('Plant breeding in an era of privatization – reflections on transformations in the Indian seed industry') reviews the status of seed system development in the light of recent experience with liberalization in the Indian seed industry. Rangnekar points out specific Indian agricultural practices while, however, contextualizing this to the global development. Referring to the Plant Breeders' Rights (PBR), Rangnekar reflects on the interactions between innovation paths and appropriation strategies in a privatized seed industry. The author characterizes 'seeds' as that instrument which captures a dual role of delivering technical change to agriculture and providing the platform for integrating farm input markets. Key indicators such as R&D investments, the number of varieties and market structure are examined in the context of a recent transformation of the seed industry in India. Rangnekar concludes with several policy implications: private sector cannot be a lone player. Few crops might attract private investment. Moreover, there is a likelihood of adverse distributional and allocative implications of private investment in agricultural research. Rangnekar makes a case for continued public plant breeding. Subsequently, he argues, the public program can consider a varietal release policy of the technology developed – offering either a free public access or else, a phased fee-based access policy, the latter holding prospects of cost recovery.

Entrepreneurship and institution

Dipendra Sinha in Chapter 8 looks at 'The rise of entrepreneurs in India'. Sinha locates certain uniqueness in the evolution of entrepreneurship in India and ponders about its future. India has a long history of entrepreneurship. After independence, India embarked upon a path of planned economic development and during this period never lacked entrepreneurs. However, after independence, the government control over various sectors of the economy grew gradually and entrepreneurs faced a number of roadblocks in the form of various regulations and licences. The economic reforms that were started in 1991 were a watershed in the history of economic events of India. While the pressures from the international agencies such as the World Bank and the International Monetary Fund were catalysts for the reforms, Sinha observes that, as conjectured by some commentators, the domestic business community as an internal voice also played an important role. This emergent voice of entrepreneurship is the concern of Sinha. Recent economic reforms have brought profound changes and opportunities for new entrepreneurs exist more than ever before. He points out that India has recently seen the emergence of a new class of entrepreneurs in the 1990s. Women entrepreneurs in particular have assumed a more significant role.

Gopinath and Vasavada in Chapter 9 ('Angels, venture capital and entrepreneurship') bring out institutional aspects of venture-support and nurturing of entrepreneurship. They discuss the role of venture capital in promoting entrepreneurship. The role of venture capitalists (VCs) in the USA is analysed, with specific focus on promoting the technology sector. A mini-case study of the Indus Entrepreneurs as a precursor network that assists in venture creation with or without capital contribution, and the institutional and societal requirements for the model to function is presented. Gopinath and Vasavada point out the importance of looking at VCs not as merely a source of financing but as a critical player within the larger context of entrepreneurship. They thus offer a critique of the short-sighted perspective in which angel/venture mechanism is looked at as a simple mechanism of financing and risk redistribution and they offer policy perspectives on venture institutions.

Knowledge economy: intangibles and labour

Facets of the institutional setting of Indian knowledge economy are brought out in Chapter 10 by Raj Aggarwal ('Business strategy and corporate restructuring in India: mutually reinforcing impact of economic liberalization and technology'). Aggarwal argues that, as in many developing countries, the Indian business environment has been much less than perfect and the inefficiency of markets for many corporate inputs has meant that Indian companies found it more efficient to internalize many of these markets. Saving upon transaction cost and the power to adjudicate on resources allocation helped the formation of Indian business houses. Consequently, most large Indian businesses have been widely diversified and vertically integrated group structures. However, the Indian economy faces accelerating change as economic deregulation and adoption of internet technology reinforce each other. In this process, Indian markets for corporate inputs are becoming more efficient and the advantages of vertical integration and diversification are declining and may even disappear. External factor markets would thus put pressure on the erstwhile internal market belonging to business houses. These changes are forcing significant adjustments in the business structure of Indian corporate. Sudden restructuring and spates of merger/de-mergers divestitures among the large and major conglomerates in India are attributable to this change. Aggarwal offers an analysis of these changes and their public policy and strategic implications.

Anupama Sreevastava and Seema Gahlaut in Chapter 11 ('Intangibles, technology trade and India: challenges before a knowledge-based economy') argue that the integration of markets and ideas, and the digital and electronic revolutions, have enabled transnational exchange of goods, services and information through intangible networks. Sreevastava and Gahlaut refer to the new phase when governments are faced with new challenges for regulating these networks, whether for patent security or national security, and they are effectively clamping down on transfer of technologies through trades in intangibles. India has carved a growing niche for itself in the global IT and space markets. It also has significant capabilities in civilian and military spheres of nuclear, missile and chemical technologies. Multilateral export control regimes control much of intangibles trade in dual-use technologies. India encouraging greater private sector participation in development of dual-use technologies implies the need to establish stronger and clearer regulations regarding technologies in general and intangible technology transfers in particular. Sreevastava and Gahlaut build upon the current perspectives and policies of the technological leaders – the United States and Europe – to identify the current perspectives and policies of the Indian government and industry. They provide specific recommendations for Indian policies with reference to this emerging area of international concern. Transparency and information sharing regarding the operations within this framework would facilitate greater technological collaboration and industrial cooperation with West Europe as well as the USA. This might also facilitate greater investment of technology-embedded capital, as well as commercial collaboration, from the advanced industrial countries. Until such time as India remains outside of the various multilateral export control agreements and arrangements, such a regulatory framework might also facilitate negotiations for pragmatic bilateral solutions that provide for greater cooperation with India.

C.S. Venkata Ratnam in Chapter 12 ('Industrial relations and increasing globalization – case study of India') discusses the labour institution of the country. He begins with a description of some of the key issues and trends in industrial relations in the wake of globalization. The institutional setting as formed through the historical development of industrial relations and the corresponding legal framework in the country provides an opportunity to examine the impact of liberalization on industrial relations in the context of increasing globalization. Venkata Ratnam picks up issues of employment, employment security and social safety nets in the context of demands from freer entry of multinationals, privatization and formation of export processing zones. This has resulted in new approaches to work organization, and application of international labour standards. Recent trends and developments in collective bargaining, workers' participation, labour laws, and new industrial relations are the results of such developments. Venkata Ratnam argues that developing a new system of industrial relations is now called for and he finally discusses how trade unions, employers' organizations and government can co-develop a new system of industrial relations.

Values and globalization

Dipak Basu in Chapter 13 ('Globalization, national culture and the future of India') brings out foundational issues to the development of new institutions in the country. Basu argues that 'globalization' stands for ultimate rights of the multinational companies to allocate resources according to their own criteria of efficiency, which may or may not correspond to the national culture of the countries in which they are operating. Efficiency, however, is a value-loaded term and a rational explication which is simultaneously acceptable to several cultures appears unattainable. Basu argues that an economic system can be successful if it corresponds to the culture or the philosophy of life of the host country, otherwise it will die out sooner or later.

Such an acceptance alone provides a foundation to the development of new institutions, new norms and new rules. In this context Basu examines whether the contemporary 'globalization' process corresponds to the basic philosophy of life of the people. He compares such global and purportedly global values with the universal values, the latter allowing full scope to local prosperity. An analysis of the Indian philosophy, as especially explained by Vivekananda, and the universal laws of nature suggest an alternative view of the world economic system. India's future will be determined by the interactions of her economic system and inherent characteristics of her national culture.

S.K. Chakraborty in Chapter 14 ('Psycho-pathology of globalization and the Indian surrender') critically examines the ideology of globalization. He raises several questions on what has been marketed vigorously over the whole decade of 1990s regarding the consumer-citizen of India, an entrepreneur-citizen of India, and whether all this made the common citizen feel secure and optimistic. He enumerates a few direct consequences of globalization upon the common citizen of a country like India. Chakraborty analyses the psychopathology of the globalizers in terms of fear, greed, addiction, ignorance and fear. Such psychic qualities, he argues, can be better described through the 'guna-triune' analytical scheme of Indian theories of mind. Chakraborty argues that globalization must now answer two tests: material sustainability and psychological sustainability. An examination of several recent international agreements on globalization reveal how miserably these fail such tests. Things of intellect must transcend borders of homes but the homes must not be swept aside – economies need not become global while universalisation must be put as a test to the world of ideas. Chakraborty, borrowing from Aurobindo, charts out the global rejections of globalization.

Bhanoji Rao and M.V. Lakshmi in Chapter 15 ('Freedom and core values for Indian development') provide a short introduction to the constituents of development in the first part. In the second part, their discussion shifts to a brief examination of the development record of India and China. They argue that neither China nor India are free from corruption and violence. The third part argues that the concept of development should be broadened to include the core values of truth and non-violence in order that the true goal of development becomes the enrichment of the human being. Rao and Lakshmi argue that development must comprise the trinity of human capability, freedom and movement towards human perfection and it must allow humanity to lead happy and contented lives and not mere happy and contented moments.

References

Arrow, K.J. (2000) 'Knowledge as a Factor of Production', in B. Pleskovic and J.E. Stiglitz (eds), *Annual World Bank Conference on Development Economics, 1999*. Washington: The World Bank, pp. 15–20.

Bassanini, A., Scarpetta, S. and Visco, I. (2000). 'Knowledge, Technology and Economic Growth: Recent Evidence from OECD Countries', *Economics Department Working Papers*, no. 259. Paris: OECD.

Chang, B.W. (2001) 'Political Economy of Intangible Investment and Weightlessness', in P. Banerjee and F.J.Richter (eds), *Intangibles in Competition and Cooperation: Euro-Asian Perspectives*. Houndmills: Palgrave Macmillan, pp. 259–75.

Chung, S. (2001) 'Knowledge Diffusion in Korean Society', in P. Banerjee and F.J. Machmillan Richter (eds), *Intangibles in Competition and Cooperation: Euro-Asian Perspectives*. Houndmills: Palgrave Macmillan, pp. 239–58.

INFAC Data Base. Mumbai: CRISIL-INFAC.

OECD (2001) *OECD Communications Outlook*. Paris: OECD.

OECD (2001) *Innovative People: Mobility of Skilled Personnel in National Innovation Systems*. Paris: OECD.

OECD (2001) *Understanding the Digital Divide*. Paris: OECD.

OECD (2000) *OECD Information Technology Outlook*. Paris: OECD.

OECD. *A New Economy? The Changing Role of Innovation and Information Technology in Growth* (Summary) at http://www.oecd.org/dsti/sti/stat-ana/prod/growth.htm

Rajan, J. (2001) 'Education in Delhi Vision 2005' (mimeo) personal communication.

Roeger, W. (2001) 'The Contribution of Information and Communication Technologies to Growth in Europe and the US: A Macroeconomic Analysis', *Economic Papers No. 147. European Commission*. Directorate-General for Economic and Financial Affairs, Brussels: European Commission.

Sarangapani, P.M. (2001) 'Indigenising Curriculum: Questions Posed by Baiga Vidya' (mimeo).

1
India's High-Tech Sector Growth: Creating a Knowledge Economy

Neslihan Aydogan

Introduction

Along with the other East-Asian countries such as China, India has recently been catching up with the global high-tech race. In fact, with highly-skilled employees, an ability to attract foreign direct investment and a series of post-1991 restructuring reforms, India seems to be on a fast track to high-tech evolution. The Indian economy gets its fuel through the conscious attention of its policy-makers, as well as the efforts of its well-educated scientists and engineers. For an economy that has been governed by protectionism and heavy bureaucracy for so long, recent developments in India are quite impressive.

Continued success for building and sustaining the high-tech sector development depends on various factors. Among these, the development of venture and entrepreneurial capital, the growth of a skilled labour force as well as the efficient organization of these factors are the most crucial.

Technology policy in India has been focused on obtaining technology by attracting foreign investors. However, there has been hardly any conscious effort directed towards transferring and adopting the key skills to organize production and exchange. Unlike firms in developed nations such as United States and Japan that have expertise in organizing production and have been successful in adapting to the specific needs of the high-tech race, India lacks such advantages. Obviously most of these skills are learned over time as firms gain experience in international markets. For countries such as India some of this knowledge can be gained through the experiences of the developed world.

For a country that is in the development process, there are other requirements for progress. Among these, a supportive infrastructure such as roads, water and electricity supplies, waste management facilities and strong financial markets are critical. For example, for the software industry cluster around Bangalore in south India, many infrastructure issues have become apparent, despite the high productivity rates in the region. As a remedy to this situation, the development of a future Industrial Park is planned to fix many of these problems in Bangalore (Singh, 1996). However, poor infrastructure remains a problem for the rest of the nation and without attention it is hard to imagine widespread growth in the most orthodox sense of the word. For example, in his seminal work, Alfred Chandler (1977) describes the revolution in transportation and communications infrastructure and the development of financial markets as an integral part of the industrial revolution in the United States.

However, the high-tech race of the recent years might be imposing different rules on firm strategies in comparison to the ones which have governed the development of the modern business enterprise in the United States. The precedence for high-tech development is likely to rest on factors that induce firms to sustain high levels of innovation. Success in organizing entrepreneurial and knowledge capital might, therefore, come prior to the development of communication, transportation and financial infrastructures. Such an argument would predict that the inadequacy in infrastructure is likely to be fixed as firms master the art of private markets. As Teece of (1996) explains: '... formal and informal structures of the firms, as well as the network of external linkages that they possess, has an important bearing on the strength as well as the kind of economic activity conducted by private enterprise economies'. Obviously it is not possible to offer general guidelines for establishing efficient inter- and intra-firm organization which would be applicable to all nations. However, a systematic study of these factors is critical for providing some important guidelines to the developing world.

Surprisingly, economists in the developed world have only recently been tackling the organizational issues related to innovation. Economists have recognized that an integral part to this analysis is the geographical space within which firms operate. Such realization is tightly linked to the sustained inequalities across regions within and across nations. In some regions both the productivity and R&D rates are much higher compared to the others. If knowledge that is produced in the United States or Japan were to spillover across different nations, the gap among different regions would effortlessly decrease. As Adams and Jaffe (1996) claim '... for a given bit of knowledge to be used it must be effectively transferred across institutional, organizational and geographic boundaries'.

Therefore firm organization and geography are interdependent and a joint analysis is necessary for understanding economic growth.

Regions of various scales, ranging from industry clusters to cities and nations that are successful in finding ways to achieve such efficiency also exhibit strengths for sustained growth. For example, as it is often noted high productivity rates in Silicon Valley rest heavily on firms' abilities to form efficient networks particularly with other area firms. Integral to such organizational networks that sustain productivity is the re-production of venture capital capabilities that flow from ex-entrepreneurs and ex-employees to start-ups. This is possible if a regional network generates constant innovation that feeds the knowledge flow across firms that belong to the network.

India is quite new in launching the primary institutions of the private enterprise. As Pandey (1998) reports, TIDICI (Technology Development and Information Company of India), the largest venture capital firm in India, has gone through an extensive learning process for organizing its supply. For high-tech technology firms that are required to compete in markets where product life-cycles are extremely short, expertise in such organization is critical.

An integral part of organizational efficiency is the requirement of a strong judicial structure for patenting innovations. Obviously, allowing firms to exercise monopoly rights over their innovations is a critical element. However, this alone does not suffice for sustained growth. A critical part of R&D activity involves knowledge that is not patentable, its exchange is costly and subject to several inefficiencies which require institutions other than courts to set the rules for its exchange. Some of these institutions are path-dependent and subject to the laws of a particular country. Some also exhibit the characteristics of knowledge exchange whether knowledge is embedded in tools, machinery, human capital or ways of doing things, and its transfer requires teaching and training and is costly. Such observation shows the importance of studying firm organization in connection with geographical location.

In this chapter my goal is to lay out the findings on the organizational aspects of innovation in relation to regions and briefly indicate the significance of adapting these to India.

Knowledge exchange, organizational efficiency, labour market fluidity and regions

Starting with the seminal work of Romer (1986), economists have been occupied in investigating the non-convexities (increasing returns) that are induced by knowledge production. Such interest goes far back in the literature, for example, Arrow (1962). The recent revival of the topic is largely related to the persistent development gaps across different regions. Obviously the sophistication of data collection and processing along with the theoretical tools that are available to economists today also help for such revival of interest.

The characteristics of knowledge and its exchange is being considered as the primary element that drives the wedge between the developed and developing world and has become a primary object of theoretical and empirical work. One major inquiry on this subject area is to find out whether knowledge is a non-rivalrous (a rival good has the property that its usage by one person precludes its use by another) and an excludable (the owner of this good can prevent others from using it) good? Or rather, is it a good which freely spills over unless the owner is granted the rights to exclude others?

Subsequent to this inquiry economists have focused on reducing the scale of analysis to firms, and classifying knowledge based on its exchange characteristics as tacit versus codified. Of course such definition is not new and in fact it goes as far back as Polanyi (1962); however, only recently have the exchange characteristics of these two different types of goods been realized as important. Coded knowledge is described as both rivalrous and non-excludable such as a computer program. Tacit knowledge, on the other hand, is described as being embodied in human capital and is both rivalrous and excludable. As Romer (1986) and Nelson (1987) describe, such knowledge can be exemplified as skills that an individual possesses, its exchange is costly and it cannot be verified by third parties such as courts. In fact these characteristics are precisely the ones that induce this type of knowledge to be geographically bounded and not spillover continents.

For example, Adams and Jaffe (1996, p. 700) find that: '...the effects of parent firm R&D on plant level productivity are diminished by both the geographical and technological distance between the research labs and plants'. This puzzle, also observed by others, is explained by Marshall (1890) in his seminal work. *Inter alia*, Marshall predicts one benefit from concentrating economic activity as being the economies of communication for transferring skills across agents. However, Marshall's definition lacks an explanation on the mechanics of such a possible occurrence.

What indeed might relate concentration of economic activity with skill transfer across firm boundaries? If skills do not carry the public good characteristics then their exchange would need to be organized via markets or by contracting between separate agents, or alternatively within the boundaries of a single firm. Therefore, learning across firms would require firms to organize such exchange.

In a parallel work, Arora (1996) describes the difficulties related to market exchange and contracting for tacit knowledge. The author claims that such exchange is subject to several inefficiencies related to the nature of such knowledge. Primarily, he hypothesizes that contracts have weak enforceability once such knowledge gets exchanged across firms. This is because the outcomes of activities in technology contracts are difficult to specify *ex ante* and verify *ex post* and hence third-party enforcement is weak. For example, the transferrer might choose to send his less skilled employees to the joint R&D lab. In addition, partner activities are difficult to monitor and hence the observability of efforts are weak. This, coupled with the uncertainty of the R&D process, might make it impossible to specify the cheating party in a contract. Given these characteristics and applying data on Indian high-technology firms Arora argues that one remedy to such a situation is to bundle tacit knowledge with something that is enforceable such as patents. This might be one way to tackle the efficiency problem, albeit it is hard to assume that such a remedy would be applicable to every transaction. Some authors, such as Morasch (1995) suggest a framework where a royalty scheme induces the licensees and licensors of technology to comply with the specifications of the contract.

Urban economists also concentrate on clusters by expanding Marshall's prediction, claiming the existence of two types of non-convexities related to the concentration of economic activity across space. One type, describes what are often called urbanization economies, economies obtained from the co-agglomeration of agents from separate industries and external to the industry in question. Localization economies, on the other hand, are economies obtained by being located in a dense area where firms from a single industry are concentrated. The empirical literature on this subject is not decisive as to whether diversity or the concentration of economic activity helps skill exchange and boosts innovation. The weakness of this literature is its alienation to the organization of such exchange, which obviously needs to be established as the crux of such an analysis.

Having established the significance of organizational choice and embedding firms inside a region, the next step of analysis requires that these elements be tied together. One such possible paradigm for such a task is transaction-costs economics. The transaction-costs paradigm (Williamson, 1984) rests on comparing the relative costs of different organizational forms. In his recent work, Williamson (1991) argues that when asset specificity is significant, it is cheaper to organize exchange inside a firm which is followed by inter-firm contracting and market exchange. This is because the tendency to cheat when exchanging specific assets is lowest when parties to an agreement are more interdependent as, for example, in the case of inter-firm contracting versus market exchange. In the above-mentioned context, knowledge can be described as a specific asset. Once the governance costs of transactions are considered – such as bureaucratic, fixed or contract writing costs – transaction-costs theory prescribes organizing exchange via markets as being the most efficient. Market exchange is hypothesized to be followed by inter-firm contracts and organizing exchange within an individual firm.

Inside regions where firms establish intertemporal networks of various forms, knowledge is likely to be exchanged efficiently without a need for transacting parties to engage in costly integration. The term network describes ranges of inter-firm contracting agreements such as supplier – buyer networks, or the turnover of skilled personnel across firms or informal contacts which include exchanging employees, machinery and ideas.

There exist many successful regions, which are all initiated with similar conditions such as the existence of an educated and skilled labour force, universities and R&D labs. It is not, however, a common trait that all such regions with all the right ingredients to begin are able to sustain growth and show increased levels of innovation overtime. As it might be apparent from the above discussion, network-building implies repeated interaction across firms inside a locale which feeds itself over time by the constancy of knowledge exchange. For example, in Silicon Valley, entrepreneurial capital emerged initially from the local universities. In addition, government injection of venture capital helped start-ups. This, subsequently, fed itself by the continuous success of firms where ex-entrepreneurs and local institutions started funding start-ups which enabled the region to be dependent almost solely on private funds. Within this structure there are two types of activities that have maintained such a turnover of funds – labour market fluidity and repeated explicit and implicit contracting among the area's firms. The first helped skilled employees to transfer skills from one firm to another. The second helped trust to be built amongst contracting firms where cheating is eliminated and knowledge is exchanged effectively across firm boundaries. Of course, coded knowledge exchange is an integral part of this system. For example, Silicon Valley engineers often boast

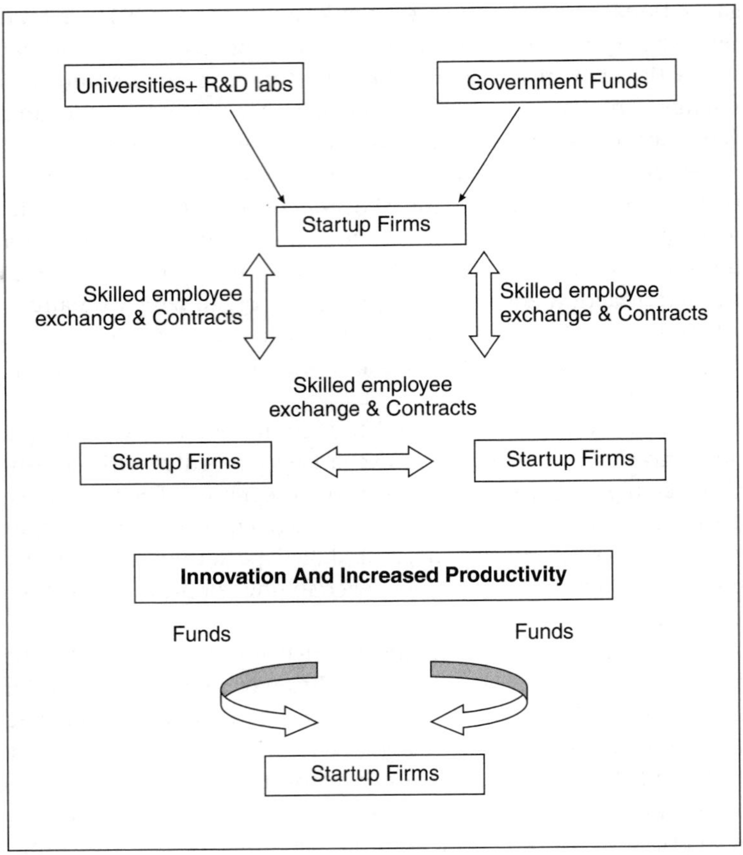

Figure 1.1 Sustained regional development with efficient knowledge exchange

about the benefits of an open atmosphere on the Valley's high productivity rates. The workings of the described system can be depicted in Figure 1.1.

Implications for the new Indian economy

The future development of the Indian economy depends largely on the Indian firms' ability to structure a system such as the one described in the previous section. Up till now, foreign direct investment has been the source where technology is imported and skilled employees of the local economy have played their role in attracting foreign firms to the region. Government, on the other hand, has been the visible hand for feeding the local firms to operate in private markets. It is highly unlikely that such a pattern could resolve itself into a high-tech economy. Whether firms are located in the software capital of India, Bangalore, or a relatively low-tech region, the rules of the game are the same. It is significant that high-tech firms in India adopt an efficient structure where firms exchange knowledge effectively. This depends largely on understanding the microstructure of Indian firms so that organizational arrangements are congruous to cultural elements. Once such a system is adopted, domestic firms' R&D output would increase and external dependency would cease. In fact, with its remarkable potential of skilled employees, the fixed costs of sustaining such a system are already reduced for the Indian economy. Now, the task is to learn the methods of private markets.

References

Adams, J.D. and Jaffe, A.B. (1996) 'Bounding the Effects of R&D: An Investigation using Matched Establishment-firm Data', *Rand Journal of Economics*, vol. 27 (4), pp. 700–21.

Arora, A. (1996) 'Contracting for Tacit Knowledge: The Provision of Technical Services in Technology Licensing Contracts', *Journal of Development Economics*, vol. 50 (2), pp. 233–56.

Arrow, K. (1962) 'The Economic Implications of Learning by Doing', *American Economic Review*.

Chandler, A.D. (1970 [c 1962]) *Strategy and Structure: Chapters in the History of the Industrial Enterprise*. Cambridge, Mass.: MIT Press.

Marshall, A. (1890) *The Principles of Economics*, vol. 3. London: Macmillan.

Morasch, K. (1995) 'Moral Hazard and Optimal Contract Form for R&D Cooperation', *Journal of Economic Behavior and Organization*, vol. 28 (1), pp. 63–78.

Nelson, R. (1998) *The Competitive Challenge: Strategies for Industrial Innovation and Renewal* (ed.) D.J. Teece. Cambridge, Mass.: Ballinger.

Romer, P.M. (1986) 'Increasing Returns and Long-Run Growth', *Journal of Political Economy*, vol. 18 (4), pp. 1002–37.

Pandey, I.M. (1998) 'The Process of Developing Venture Capital in India', *Technovation*, vol. 18 (4), pp. 253–61.

Polanyi, M. (1962) *Personal Knowledge*. London: Routledge & Kegan Paul.

Roderick, M. and Chandramouli, M.S. (eds) 1999 *Doing Business with India*. Price WaterHouse Coopers, (1996) p. 139.

Singh, A. (1996) 'Karnataka: India's High-tech Eden', *Director*, vol. 50, (4), p. 23.

Teece, D.J. 'Firm Organization, Industrial Structure, and Technological Innovation', *Journal of Economic Behavior and Organization*, vol. 31, (2), pp. 193–224.

Williamson, O.E. (1991) 'Comparative Economic Organization: The Analysis of Discrete Structural Alternatives', *Administrative Science Quarterly*, vol. 36 (2), pp. 269–96.

Williamson, O.E. (1987) 'Transaction Cost Economics: The Comparative Contracting Perspective', *Journal of Economic Behavior and Organization*, vol. 8 (4), pp. 617–25.

2

Information Technology as an Engine of Broad-Based Growth in India

*Nirvikar Singh**

Introduction

Nothing has captured the imagination of India's policy-makers quite like information technology (IT), and Indians have proved themselves to be world-class in IT. Indians (or people of Indian origin) are becoming not just contributors but leaders of the IT revolution in the United States, and India's software industry has appeared to be growing so rapidly that it would forever solve India's balance of payments problems, and make India an IT 'superpower'.

At the same time, some have decried India's IT boom as just another version of global sweatshop production, with lines of code replacing garments. A related concern is that the growth will peter out as these labour-intensive software tasks reach their limits, or even become automated. More worrying, perhaps, is the concern that IT in India will become an enclave similar to natural-resource enclaves, controlled directly or indirectly by multinationals with a small elite of workers, no productive spillovers to the local economy, and greatly increased inequality. Some of the different concerns are mutually exclusive, but either possibility, rapidly diminishing returns or very narrowly focused benefits, is inconsistent with a golden future for India's IT sector and its broader economy.

* In preparing this chapter, I was grateful to Yale Braunstein, Ashok Desai, Atanu Dey, Rafiq Dossani, Kyle Eischen, P.D. Kaushik and Vini Mahajan for helpful comments and discussions. None of them is responsible for any remaining shortcomings. I have learned about aspects of this topic from numerous people in addition to the ones named; I have tried to reference their works as much as possible, and regret any inadvertent omissions.

In this chapter we survey some of the developments in India's IT sector, and prospects for broad-based growth led by this sector. We try to identify some areas where policy changes and current innovations can together lead to realization of the more optimistic scenario, and avoidance of some of the pitfalls that analysts have identified. In the next section a brief overview of the Indian economy is provided, including its structure and recent reforms. I then examine the IT sector, discussing the role of software versus hardware, the growth pattern of the software industry and software exports, and the potential problems in IT labour supply to support future growth. The current bottleneck for the IT sector, namely the telecommunications infrastructure, is examined, and issues considered include the basic driver of technological convergence across voice and data communications, problems with current infrastructure, innovations that have the potential to dramatically alter the economics of access to telecoms, and the evolving structure of the telecoms industry.

While policy issues crop up throughout the chapter, the policy environment in particular is examined more closely, arguing that government policy is better focused on removing labour-market distortions and infrastructure constraints, rather than providing output or export subsidies to the software industry. The latter are not only costly, but will lead to narrower development of the IT sector. I discuss the appropriateness of specific policy goals such as universal access, as well as issues of implementation of more general objectives of broader telecoms access.

Building on earlier arguments in the chapter, the possibilities for broad-based IT-led growth are mapped out, including increasing value-added, using better telecom links to capture more benefits domestically through offshore development for industrial country firms, greater spillovers to the local economy, broadening the IT industry with production of telecom access devices, improving the functioning of the economy through a more extensive and denser communications network, and improving governance. The final section provides a summary conclusion.

Economy overview

India has made considerable economic progress since independence, but it has not grown as rapidly as the East Asian 'miracle' economies. Economic policy emphasized state-led industrialization until the 1990s, though there was some shift away from this in the 1980s. Growth accelerated somewhat in the 1980s, and even more in the 1990s, in which GDP growth averaged around 6 per cent. The recent sectoral distribution of GDP is shown in Table 2.1, which also illustrates that the level of

Table 2.1 India's GDP (Rs. billion) at factor cost by industry of origin, 1993–94 to 1999–2000 at 1993–94 prices

Year	Agriculture, forestry and logging, fishing, mining and quarrying	Manufacturing, construction, electricity, gas and water supply	Transport, communication and trade	Banking and insurance, real estate, dwellings and business services	Public administration and defence and other services	Gross domestic product at factor cost
1993–94	2620.6 (33.5)	1850.7 (23.7)	1505.0 (19.3)	900.8 (11.5)	936.3 (12.0)	7813.5
1994–95	2760.5 (32.9)	2040.9 (24.4)	1661.3 (19.8)	950.9 (11.3)	966.7 (11.5)	8380.3
1995–96	2751.5 (30.6)	2291.0 (25.5)	1881.7 (20.9)	1028.5 (11.4)	1043.0 (11.6)	8995.6
1996–97	2994.6 (30.9)	2468.5 (25.4)	2029.4 (20.9)	1099.9 (11.3)	1108.4 (11.4)	9700.8
1997–98	2950.5 (29.0)	2561.1 (25.2)	2185.1 (21.5)	1227.8 (12.1)	1238.2 (12.2)	10162.7
1998–99	3144.0 (29.0)	2654.3 (24.5)	2340.2 (21.6)	1331.3 (12.3)	1360.7 (12.6)	10830.5
1999–2000	3167.8 (27.5)	2837.2 (24.6)	2528.3 (21.9)	1465.5 (12.7)	1521.2 (13.2)	11519.9

Notes: 1999–2000 figures are estimates. Numbers in parentheses are percentages of total GDP.
Source: http://www.finmin.nic.in/fecosur.htm

GDP remains quite low at about $300 per capita. Poverty rates have fallen gradually over time, and somewhat more rapidly in the last decade, but they remain quite high, at close to 30 per cent of the population. Human development indicators such as literacy and life expectancy have also tended to lag behind targets, with adult literacy being only 65 per cent according to the latest census (even with a very minimal criterion being applied).

The table illustrates that agriculture remains an important part of the economy, contributing close to 30 per cent of GDP and still ahead of manufacturing. Agriculture's share of employment is considerably higher. Services of all kinds and the public sector are the other two sectors of the economy. While the public sector is, by its nature, 'organized', much of the service sector, like agriculture, is 'informal' in nature. Thus India preserves the classic 'dual' nature of a developing economy.

The single most important change that was introduced in the early 1990s was a drastic paring down of the longstanding regime of government controls on private industrial investment. The combination of this relaxation of the 'licence-permit *raj*' and substantial reductions in trade barriers (as well as moving from quantitative restrictions to tariffs) arguably contributed to the faster growth that India has seen in the past decade. Foreign investment was also liberalized. Liberalization of previously very severe controls was easier to accomplish, in some respects, than deeper institutional reforms, which include privatization of parts of the economy that have had substantial government presence, and the development of a set of modern laws and regulatory institutions to provide a framework within which the private sector can operate effectively.

Both political obstacles in the form of vested interests, and difficulties in learning how to create and manage new institutions in an environment of simultaneous rapid globalization and technological change, have slowed the progress of fundamental reforms of laws and institutions. Some of the greatest difficulties have arisen in sectors where private enterprise has traditionally been viewed as subject to market failure problems, due to cost structures that supported natural monopolies (that is, large scale economies), or the existence of externalities or public good characteristics (that is, non-rivalry or shareability, and non-excludability). The technical difficulties of efficient economic organization of production in such areas have been compounded by political resistance to change, even though the existing organization is grossly inefficient.

Two key sectors that stand out in this respect are electric power and telecommunications (telecoms). Along with roads and ports, power and telecoms are the essential infrastructure for any economy, and they represent the worst bottlenecks that currently prevent India from achieving higher growth rates. If one has to pick one of these sectors, electric power is the area where the problems and the benefits to improvement are arguably the greatest, but in this chapter we take this as given, and focus on telecoms in the context of information technology more generally.

The IT sector

Information technology essentially refers to the digital processing, storage and communication of information of all kinds.[1] Therefore, information technology (IT) can potentially be used in every sector of the economy. The true impact of IT on growth and productivity continues to be a matter of debate, even in the United States which has been the leader and largest adopter of IT. However, there is no doubt that the IT sector has been a dynamic one in many developed countries, and India has stood out as a developing country where IT, in the guise of software exports, has grown dramatically despite the country's relatively low level of income and development.

Software vs hardware

The basic distinction in IT is between hardware and software. The former refers, of course, to the physical components of processors, storage devices and communications devices. The latter refers to the instructions that govern the flow and processing of information in digital form within and between hardware devices and components.[2] The production of hardware is classified within the manufacturing sector. Profitably manufacturing semiconductors and other sophisticated hardware components typically requires infrastructure, large-scale investments in capacity, and accumulated experience that India does not possess, and is not in a position to acquire easily. India's development path, despite its emphasis on import-substituting industrialization, has not supported the growth of a robust, world-class manufacturing industry such as has arisen in many East Asian countries.

Nevertheless, India does perform many hardware assembly tasks internally for the domestic market; components in such cases typically come from East and Southeast Asia. The ability to organize this aspect of production may be the basis for further development of hardware capabilities. Several East Asian countries also began as mainly assemblers of sophisticated components produced elsewhere, and extended their presence in the value chain backwards as they learned by doing. While being late to the game may make entry more difficult, the fact that manufacturing of components becomes increasingly standardized, and the cost of these components falls, works in favour of late entrants. For example, the production of most memory chips has become commoditized and moved to developing countries, where 20 years ago it was the core of Intel's business.

The example of firms like Dell and Cisco may also be useful to keep in mind when evaluating the hardware industry in India. Dell outsources most, if not all, of its component manufacturing. It is, in fact, an extremely sophisticated assembler, and its value creation is based on organizing this assembly as efficiently as possible, doing so on demand and keeping its inventories absolutely minimal. Strong customer service plus management of communications and logistics at both ends of the value chain are also keys to Dell's success. Dell's positioning to take advantage of strengths in infrastructure and closeness to a growing customer market is an important lesson for India. The point here, which we shall further develop subsequently, is that a hardware industry in India is feasible, building on India's experience in assembly, but it will require significant improvements in infrastructure, and careful attention to market needs.

India's software industry is more robust than its hardware industry, at least in certain areas. While selling packaged software to consumer (and most business) markets requires economies of scale and scope, as well as marketing and customer support muscle, project-oriented components of software development do not do so to the same degree. The software development and use life-cycle includes analysis and specification of requirements, design, coding, testing, installation, maintenance and support. Many of these activities, particularly coding and testing, involve *relatively* routine IT skills that India's workforce has in large absolute numbers (though small relative to the total population).

The existence of the Indian Institutes of Technology (IITs), the ubiquity of Unix in academic environments, and the relatively low infrastructure demands of learning to use and create software all worked in India's favour on the supply side. The use of English in India's higher education system, the increase in the use of Unix and related operating systems due to the explosion of the Internet, and the large number of Y2K-related projects in the late 1990s all contributed to demand for India's software industry services, in addition to the general growth in

IT in the 1990s. Much of this demand came from abroad, as we discuss in the next subsection. However, the software industry's domestic revenues also grew rapidly in the last few years, at over 30 per cent per annum, on average.

Despite the even faster growth of software exports, domestic software revenue still represents close to one-third of software industry gross receipts. The National Association of Software and Services Companies (NASSCOM) projects domestic sales to grow substantially faster than export sales in the next decade, enough to make domestic sales over 50 per cent of the industry's sales, but the basis for this projection is unclear. In any case, India's software industry is quite robust. In the domestic market, products and packages make up almost half of revenues, with projects accounting for over a quarter, while professional services, support and maintenance, IT-enabled services and training each make up less than 10 per cent.[3] This pattern is quite different from that of exports, as we discuss below.

Software exports

India's software exports, in particular, are what have captured the headlines. A growth rate in software exports of over 50 per cent (albeit from a low base) for several years, and consulting firm McKinsey's projection that software export revenue would reach $87 billion in a decade, are two of the most striking statistics in this regard. The latter figure would represent over half of India's projected payments on current account in a decade.[4] More conservative export growth projections used by NASSCOM would still imply that software exports would account for over a third of payments for visible and invisible imports. It must be borne in mind that the current figure for software exports is closer to $6 billion.

The pattern of activities that generates software export revenue is somewhat different from the sources of domestic revenue. In particular, professional services accounted for 44 per cent of export revenue in 1998–99, followed by projects at 36 per cent. Products and packages were only 8 per cent of the total compared to nearly half for domestic revenue. Desai (2000) suggests that some of the difference in the patterns of domestic and export sales is illusory, because domestic sales of packages are by resellers of packaged software that is licensed from foreign software developers or vendors.

Coding and testing therefore appear to form a significant proportion of the skills used in the Indian software industry. The concerns expressed by analysts[5] are that the Indian software industry is 'programmer heavy', and therefore unable to move up to higher value-added segments of software. Related issues that reinforce these concerns are the brain drain of the most talented or experienced IT people, the lack of sufficient managerial skills for more sophisticated contract work, and the lack of domestic spillovers from the 'body shopping' of programmers for onsite work in developed countries.

Much of this body shopping takes place in the largest market for India's software exports, the United States. The USA is the destination of two-thirds of India's software exports, with Europe far behind at 21 per cent (half of that being in the United Kingdom), and Japan and the rest of the world accounting for the remaining 12 per cent.[6] The US share in exports has risen along with its booming economy, and the slowdown there will no doubt reduce that share somewhat, as well as the growth rate of India's software exports.

IT-enabled services

As the term suggests, IT-enabled services are not necessarily related to the production of software or IT in general, but use IT to make the provision of services possible. The figures for the software industry in India typically include these IT-enabled services, though they are not strictly part of the IT sector. Customer call centres are one example, where Indians have been training to speak with American accents in order to deal with customer queries from the USA. Accounting services are a second example. Yet another, more long-standing market segment is that of medical transcription.

The medical transcription segment illustrates some of the general problems with Indian industry serving foreign markets. Firms are often too small to market to or deal directly with clients, and relying on intermediaries reduces their margins, making investments in adequate training difficult or impossible. In turn, inadequate training hurts the quality of transcription, ultimately raising costs because problems must be fixed. Despite the success stories of Indian firms in this segment, revenues are only in the millions of dollars, whereas the US medical transcription market outsources several billion dollars worth of business worldwide.[7]

Good communications links are obviously important for the success of IT-enabled services such as medical transcription provided to foreign clients. The severest bottleneck in India, however, may not be telecoms, but the lack of managerial and marketing skills, and of reputations for quality. Recent developments suggest that this may be changing. Part of the solution includes the import of such skills by multinationals such as GE and Citigroup shifting some of their back office operations to India.

This trend is behind the optimistic projections of a report submitted to the Indian government's Electronics and Computer Software Export Promotion Council, which sees IT-enabled services exports growing from $264 million in 2000 to over $4 billion in 2005.[8] While these projections may be on the high side, even half as much growth would be impressive. We will return to issues of managerial skills and moving up the value-added chain later in the chapter.

Supply of IT skills

The reason for the success of India's software industry is the large supply of labour with some IT skills. India graduates perhaps about 125 000 engineers a year, second only to the USA worldwide.[9] However, not all these engineers go into the IT industry, and not all IT professionals have engineering or computer-science qualifications – this being true of the USA as well.[10] India's stock of IT professionals is estimated at 300 000,[11] so that software industry revenues per IT professional (assuming that all of them work in the software industry, which is unlikely) are about $30 000.[12] Government targets and others' optimistic projections imply software industry revenues will increase by a factor of 15. The breakdown of this growth could be something like a doubling in revenue per IT professional, and therefore almost an eightfold increase in numbers.

Both growth components have implications for IT training. Increasing revenue per IT professional requires improvements in managerial and marketing skills,[13] but it also requires the production of more highly trained IT people. Training more people in IT requires investments to increase the capacity of this component of the higher education sector. This is an extremely thorny problem. Even at the elite IITs, faculty are poorly paid relative to industry, and the physical infrastructure has deteriorated from lack of investment. One might argue that this could be fixed by increases in government expenditure, but this is difficult in an environment of large budget deficits and long-term neglect of basic education.

Of course increased private investment in, and operation of, IT-related training and education is an option that is rapidly developing. One potential problem is that of maintaining standards and quality. However, reputations should be possible to establish, so that well-educated IT professionals will do better on the job market over time. IT industry investment may also play a role, since the industry has a strong interest in growing the available supply of IT professionals. Private providers such as Aptech, which characterizes itself as 'Asia's leading provider of

IT training', are also playing a role. Finally, given the strong reputation and talent pool in the IITs and similar institutes, however, it does seem that government investment may be better channeled toward them, rather than indiscriminately setting up new institutes for IT training, as the Ministry of Information Technology seems to be doing.[14]

A further problem besides sheer numbers is the issue of the level of training, and even the IITs are hard pressed to provide postgraduate education comparable to what is available in the USA. Increasing the level of IT education in India may simply exacerbate the brain drain,[15] as the most qualified continue to be attracted to developed countries. Desai (2000) uses this issue to suggest that India may actually be better-off by continuing to specialize in the lower end of the market, for coding and testing, as well as in IT-enabled services, at least in the next few years. Which route is most profitable is best left up to the players, with the government's role being to avoid excessive policy distortions that create imbalances across different segments within the IT sector. In any case, the supply of skilled labour is an important consideration for the growth of the sector, both domestic and export-oriented.

Telecom infrastructure

Basic hardware and operating system software (particularly open source for the latter) have declined in terms of both absolute and relative cost. Liberalization of imports has allowed Indian IT firms to more efficiently obtain these inputs into their own production of software and services. While a shortage of skilled labour looms as a future problem, one of the most severe constraints has been the poor state of India's telecoms infrastructure.[16] The benefits of well-functioning telecommunications are much broader than just in IT, but the Internet and the associated IT boom have made India's telecoms bottleneck a greater concern. At the same time, rapid technological change implies that the ability to alleviate this bottleneck may be much greater than just a few years ago.

Convergence

'The Indian telecommunications system continues to be governed by the provisions of the Indian Telegraph Act, 1885 and the Indian Wireless Act, 1933.' This statement is from the New Telecom Policy of 1999. Until recently, the telephone was considered to be a luxury good by government policy-makers. These observations illustrate India's starting point in a world where telecommunications are being transformed incredibly rapidly.

The ability to digitally encode all kinds of information, whether voice, data or video, makes it possible to send all this information over a single network with digital capabilities. This combined network may include copper wires, fibre-optic cables and wireless transmission; this is the essence of what we mean by 'convergence'. Additional kinds of convergence are possible. The software protocols that govern communications can converge; the same devices can be used for access of different kinds (data and voice conversations, for example). However, these types of convergence are more specific, and are not necessary consequences of basic convergence of digital networks.

The implication of convergence is that telecoms are receiving more attention than in the past. While India began to encourage the setting up of Public Call Offices (PCOs) throughout the country in the 1980s, teledensity remains very low, between 2 and 4 per hundred. The quality of lines and exchanges is poor, and most telecoms remained a government monopoly until very recently, failing to follow quickly on the path of liberalization begun in 1991. It has been the rise of India's software industry that has focused attention on technological convergence, and the benefits and feasibility of dramatic change in the telecoms sector.

Infrastructure needs

The software industry uses international data links for accessing clients' hardware, communicating by e-mail, exchanging files among joint development teams, and carrying out remote diagnosis and maintenance work.[17] IT-enabled services use voice lines for call centres, and data lines for transmitting electronic files back and forth. Internet-based media companies also require data links. While all economic activity requires good communications infrastructure, the rapid rise of the Internet has directly and indirectly increased the need for such infrastructure. Meanwhile, India's traditional infrastructure remains poor, with its teledensity being well below other developing countries such as China, and a low penetration of wireless access.

The demands on the telecoms infrastructure span an enormous range. While software firms, other exporters and multinationals require robust high-bandwidth international links, the great majority of India's population has little or no access to the most basic voice services (see further for details). It may seem that the needs of this range of users are so different that they require very different solutions. This is partly true, but technological convergence, and the fact that the domestic and international networks must interconnect, imply benefits to integrated approaches to infrastructure improvements. We therefore examine the network at several levels.

International links are an obvious area for improvement if the Indian software industry is to realize its lofty growth projections. Current bandwidth is only a small fraction of what is required. A key problem has been the monopoly of the government-owned Videsh Sanchar Nigam Limited (VSNL), which has been the sole provider of international access. VSNL's monopoly has severely restricted the expansion of international links, at a time when the cost of doing so has been falling rapidly. It has also greatly hampered the growth of domestic ISPs. After prolonged criticism from numerous sources, the monopoly finally ended in April 2002.

The domestic infrastructure is equally lacking. Here the government has already introduced competition, but there have been several policy missteps which we discuss later. Lack of adequate attention to building a domestic Internet backbone has been one problem, which has contributed to the use of US-based servers for hosting 'Indian' web sites and e-mail services, even for purely domestic users, hence increasing the demands on international links and bandwidth. The paucity of the traditional voice infrastructure (both in terms of bandwidth and switching capabilities), which carries Internet traffic also, further inhibits modern communications. Internet use ties up lines for longer periods than voice calls, exacerbating the problem.

While increased and better access to the Internet is an important issue, it comes up in a situation where even the most basic telephone service is unavailable to large segments of the population. Official figures state that over 60 per cent of villages have public telephones (VPTs), but most (over 90 per cent) do not have direct long-distance dialing, and the quality and maintenance of the service is possibly quite low. Again, there are problems with the exchanges that serve these rural areas, which increasing the number of lines and access points will not solve. A problem with increasing access in rural areas is that the fixed cost of providing access is high compared to the potential revenue generation.

Many of the inadequacies of the telecoms infrastructure arise from government provision or inappropriate regulation, an approach that emphasizes quantitative targets, and a lack of coordination between developing different parts of the network. These problems are illustrated by the requirement that three new private fixed service providers (FSPs), awarded licences, install 42 841 village pay phones in their first two years of operation, without regard to the economics and enforceability of this requirement.[18] Of course, the FSPs were being explicitly required to engage in cross-subsidization, but even if successful it would result in only an extension of inadequate service, with low utility for users as long as other bottlenecks or economic constraints such as high interconnection charges are not addressed. In fact, recent technological innovations make much more than this possible, as we discuss in the next subsection.

India's telecoms infrastructure needs must be understood not as providing high-quality, high-speed international data links for its premier software firms, but independently providing basic local voice calling capabilities for poor villagers. While clearly what is provided is closely linked to ability to pay, the value of being part of a well-functioning network must not be underestimated, even for the bottom rung. For example, the average annual revenue per line of VPTs is estimated at \$16 without long distance, and \$760 with that capability.[19] The essence of a network is connectivity, and the domestic network and international gateways, fixed and wireless service, and voice and data must all be built out in a coordinated manner to maximize the value of the network to its users. The government's role should be to ensure this coordination, without stifling competition and innovation, as we shall later discuss further. We next discuss innovations that potentially provide cost-effective voice and data telecom access to rural and semi-rural populations. To the extent that such access can stimulate demand for IT products and services geared to the domestic market, there is a positive link between widespread telecom access and the domestic IT industry.[20]

Innovations

Some of the key work on innovation for Indian telecoms has been done by teams led by Ashok Jhunjhunwala, of IIT Chennai.[21] He realistically frames innovation needs in the context of economics; affordability is critical to making widespread provision of telecoms services economically viable. Jhunjhunwala gives the example of cable services in India, which are priced at \$2 to \$4 per month and have 35–40 million subscribers. At this kind of price point, however, a telecom operator in India cannot recover set-up costs for access which are about \$800 using conventional technologies.

The economics of widely available telecom services in India is therefore very different from the USA, where revenue per connection will be several hundred dollars per year: innovation in the USA focuses on increasing revenue through upgrading services, rather than reducing the cost of providing access. The goal of innovations by Jhunjhunwala's team, therefore, has been to bring the cost of access down below $300 per line, and as close to $200 as possible. The latter figure would make access affordable to 50 per cent of Indian households at current income levels. On the other hand, without such innovations, targets of increasing India's teledensity fourfold (from 4 to 15 per hundred), or Internet access tenfold are empty rhetoric.

With the cost of fibre-based backbones falling rapidly, it is the access component of the network that accounts for as much as two-thirds of the per-line cost. The IIT Chennai group and spin-offs started by alumni have developed several key innovations that can dramatically bring down the cost of access. These innovations include developments in hardware as well as software, and address issues of network management and deployment as well as pure access issues.[22] They help bring both affordable voice and Internet access to rural areas of India.

The benefits of this suite of technologies are not restricted to rural users, but also extend to middle-class and working-class urban users. Current access costs using these new technologies are estimated at $400 per line, but are likely to fall with further innovation. Pilot projects in rural and urban areas appear to have been very successful, and adoption is finally gaining some traction, despite bureaucratic and policy hurdles (see our later discussion of the policy environment).

The bottom line is that bringing down the cost of access through innovation targeted at the domestic market is a critical component of any dramatic increase in telecoms connectivity in India. Economically combining Internet and voice access also has the benefit of increasing the value of connecting to the network. The benefits accrue not just to the poor, but also to the tens of millions of lower middle class households who are currently outside the affordability radius. A denser domestic network not only increases the value of international network links, but it also provides opportunities for increasing the rate of training of IT personnel, which is a looming bottleneck (see p. 32 above). Finally, the development of an indigenous hardware industry for low-cost access devices and network components has the potential to fill in gaps in India's IT capabilities on the manufacturing front.[23] For this potential to be realized, there will have to be some important policy changes. Before we turn to the policy environment, we briefly consider the organization of some aspects of India's telecoms industry.

Industry organization

Telecommunications was a nationalized industry before 1994, when privatization began. Privatization has consisted of auctioning off spectrum rights for wireless or mobile services, and, more recently, auctioning licenses for fixed service provision. In theory, under some specific assumptions, auctions have the property that those who can provide the greatest value (and hence reap the greatest profits) would bid the highest, achieving both revenue raising and efficiency goals for the government. In practice, the theoretical assumptions required for this result may not hold.[24]

The geographical units for which licenses have been auctioned have been quite large, presumably based on the consideration of economies of scale. Thus large domestic corporations and multinationals have tended to be the successful bidders (with some notable exceptions, such as HFCL). However, even deep-pocketed successful bidders have found that net revenue generation opportunities were more limited than expected, given the technology being used and the actual market sizes.[25] As a result, upfront license fees have been replaced with revenue-sharing arrangements.

If revenue sharing is used anyway, then some of the advantages of having large firms as providers (financial strength) are attenuated. On the technology side, the work of Jhunjhunwala and his group suggests that economies of scale no longer require the service areas to be as large as states[26]. Even if there are factors such as building the backbone, which favour larger service areas, all aspects of the network do not have to be provided by the same firms. One possible example of industry organization is that of cable TV, where small operators connect to a larger network, often operating at a level as small as a group of urban neighbourhoods. Of course cable operators offer one-way, last-mile connectivity to a much larger broadcast network, and the telecoms network is somewhat more complicated in these respects. Thus the argument is not necessarily for smaller franchise areas, but for a system where tiered franchising might be possible.

A serious problem with a decentralized or franchise model of access provision is the monopoly aspect of some components of the network. A monopoly charges high prices and hence restricts output. However, even a profit-maximizing monopolist would do better to allow small access providers to connect to the backbone network at rates that allow the access provider to make a normal rate of return, than to shut them out entirely, if the small firm is more efficient in its operations. The real problem in India may be the imposition of high license fees and interconnection charges to new local providers by the biggest monopolists in the telecoms industry, namely state-owned firms. Adjusting these policies can therefore lead to an industry organization that is better positioned to take advantage of recent innovations in hardware and software for access, and more quickly extend the reach of the telecoms network.

The policy environment

Since 1991, India has pursued policies of economic liberalization, but policy reform has been uneven. Controls on private industry and non-tariff trade barriers have been removed or substantially reduced, but liberalization has been slower in areas where there is clear interest group opposition, such as labour laws and privatization. Reform has also been slow in areas where new regulatory institutions needed to be created: there is still a substantial amount of learning by doing that is taking place. Finally, government revenue considerations also affect policy decisions in areas such as import tariffs and telecoms privatization. The IT sector in India is important not just because of its performance and potential, but because these factors have influenced the policy environment in India.

Policy goals

The overall goals of policy are quite standard: high growth together with macroeconomic stability and poverty reduction. Balancing these goals is the difficult part. For example, incentives for exports, such as tax breaks, are designed to spur growth, but may adversely affect the government's fiscal deficit. As quantitative controls have receded in importance, such tax-subsidy policies have become more significant policy components. The growth of India's IT sector, and the success of the software industry in particular, has tended to skew policy toward the industry, with targeted incentives being implemented or recommended.

Targeting incentives to the software industry is not necessarily the best method to promote the industry, nor to achieve broader goals of growth and human development. Providing implicit or explicit subsidies to the industry can introduce distortions, and it involves forgoing other uses of funds, given the severe budget constraint that the government faces. Broader promotion of the IT sector also suffers from some of the same problems. Investing heavily in government-sponsored IT-related training is problematic when basic education in India is so poorly provided by the public sector. Policy goals for the IT sector would be better met by focusing on infrastructure provision, enabling the private sector to play a role here as well. The telecoms sector is a case in point.

The historical case for regulation or nationalized provision in telecoms was based on economies of scale, implying that competition would be unstable or inefficient. Technological change has removed this justification in significant portions of the telecoms value chain by lowering fixed costs and adding new technological options, allowing competition to become feasible. Since monopoly may persist in portions of the value chain (that is, portions of the network), regulation of interconnection charges may still be required, to maintain a level playing field.[27] However, directly managing technology choices and competition is not easily justified on economic grounds. The broad policy goals of promoting competition and innovation in the provision of telecoms infrastructure and services are moving in the right direction, if one compares the NTP of 1999 with that of 1994, but the proposed details of implementation leave much to desired, as we discuss in the remainder of this section.

Regulatory institutions and laws

Broad-based growth of India's IT sector will depend on improving the telecoms infrastructure, and on training enough people for the sector and using them effectively and efficiently. For telecoms, the regulatory framework is crucial, whereas for human-resource development and use, the labour laws matter greatly. It may also be noted that laws that directly constrain manufacturing remain on the statute books, and adversely affect areas such as manufacturing or assembling hardware – the problem here is one that still affects Indian manufacturing in general.[28]

In India the regulatory institution for telecoms is the Telecom Regulatory Authority of India (TRAI), which was constituted in 1997 and substantially modified in structure and role in 2000. The new TRAI has more authority than the old, and some contradictions in the 1997 legislation have been removed.[29] However, the government's Department of Telecommunications (DOT) still has final authority over licenses and license terms. The difficulties and costs of obtaining licenses for local service provision are, in fact, a major barrier to increased competition and the implementation of the innovations discussed earlier in the chapter (pp. 36–37). The DOT controls the domestic long-distance monopolist, Bharatiya Sanchar Nigam Limited (BSNL), and the present regulatory structure involves a conflict of interest.

In general, the present TRAI has a broad scope, including establishing quality of service parameters, monitoring compliance, examining technology choices, and so on.[30] It is supposed to establish a level playing field and encourage competition, but it lacks clear authority precisely where it needs it the most, in setting entry fees and some interconnection charges. Unfortunately, bringing quality of service, technology choice, and universal service obligations (see below) into the regulatory mix only serve to muddy the waters, and divert attention from the central task of enabling effective competition.

In May 2002, the central government introduced a Communications Convergence Bill, which is meant to establish a unified regulatory framework for voice and data communications. While this development is welcome, as our previous discussion suggests, it is the details of the regulatory set-up and its implementation that will matter most, rather than the recognition of technological convergence. One practical development that promises to be significant is the allowing of calls over the Internet (VoIP) from April 2002. As of August 2002, the Convergence Bill itself was still winding its way through the legislative process.

Desai (2000) examines the problems of labour laws, using the Report of the Subject Group on Knowledge-Based Industries (2000) as his starting point. The report calls for exemption for the IT sector from a broad set of rules relating to labour, including provisions relating to overtime, working conditions, restrictions on contract labour, and dismissal of workers. Interestingly, if IT workers are in short supply, they should be able to negotiate terms that are attractive enough to make the labour laws redundant. Desai suggests that the main function of the labour laws in this sector is to enable government labour inspectors to demand bribes. He also argues for broader reform of labour laws, and rightly points out the potential for distortions if one sector is given an exemption. He also acknowledges the political difficulties of more comprehensive reform. In this case, the IT sector may usefully serve as the thin edge of the wedge that begins cutting down some of the worst problems with India's labour laws, in particular the lack of permitted flexibility in contracting.

Access and efficiency

One interesting feature of recent policy pronouncements and targets has been the emphasis on increasing access to telecommunications. This contrasts greatly with previous Indian policy views of telephones as luxuries. Universal service obligations are being built into licensing deals for private service providers, in the form of quantitative targets for installing rural telephones. As we have discussed earlier, there are problems with enforcing such requirements, and in fact it is not clear that numerical targets of this sort have any use at all, especially when licensing and interconnection fees make it uneconomical for local access providers with lower-cost technology to enter.

A different issue is the problem with cross-subsidization when the user base is so small. A literal universal service obligation (USO) with cross-subsidization within the sector would require taxes on this small customer base to support providing access to new users. This model is inherently unworkable, as pointed out by Dey (2000). USOs have worked in circumstances where the user base is the great majority of the population, but imposing a USO prematurely, as Indian policy now seeks to do, makes no sense. The problem is even more striking when the contradiction between government USO rhetoric and government-imposed entry barriers is noted. Subsidizing entry by new competitors makes more sense than imposing quantitative targets that are hard to enforce meaningfully. There are similarities between the requirements imposed on new private entrants to install VPTs and the traditional mechanisms of Indian planning, which have emphasized spending money on projects without making them economically viable enough so that they actually provide a stream of benefits.

Ultimately, neither the TRAI nor the DOT will be able to effectively manage quality of provision, technology choices or true availability of access. Instead, the focus should be on actually achieving a policy environment where competition is effective. When consumers have the option to switch, this will force competitors to pay attention directly to quality of service and technological efficiency. When firms can make profits extending the network, without being blocked or impeded by government fees and restrictions, they will provide access efficiently. This scenario may have been impossible just a few years ago, but technological change has altered the possibilities, and government policies have to catch up.

IT as a growth engine

Ultimately, the case for IT as an engine of growth rests on standard economic criteria such as comparative advantage, complementarities and global dynamics. Briefly, the IT sector can be an important source of growth for India if the country has a comparative advantage in providing certain kinds of IT-related products and services, if the global demand for these products and services is likely to grow rapidly, and if the growth of the sector has positive spillover benefits to the rest of the domestic economy. Furthermore, these conditions are not purely exogenous, but are partly functions of economic policy. We examine these considerations in this section.

Value added

The static theory of international trade is based on comparative advantage, determined by relative factor endowments and/or technology differences.

Empirically testing this theory is difficult, however, since other factors may also be at work. For example, intra-industry trade driven by product differentiation and economies of scale may involve different trade patterns than those based on traditional comparative advantage. In the case of software exports, attributing the export boom to comparative advantage does seem reasonable.[31] We have noted India's pool of workers with software and language skills that are valued in the international market: they are the source of India's comparative advantage in at least some segments of the software industry.

The comparison of software exports involving coding and testing with garment exports and sweatshops, while typically made in a negative manner, actually has positive implications for India's software industry. While India missed the boat with respect to the labour-intensive manufactured exports that contributed to the East Asian miracle, it is now in a position to replicate this phenomenon with labour-intensive software services and (even more labour-intensive) IT-enabled services. Even if exports of this nature cannot sustain growth rates of 50 per cent (and the recent slowdown seems to confirm this caution), they can make a substantial contribution to India's economic growth. For example, 20 per cent growth in a sector that is only 5 per cent of the economy still adds one percentage point to overall economic growth. In the very short run, therefore, moving up the ladder of value added (or establishing a broader hold on the value chain) may not be a critical issue. This strategy also addresses brain-drain issues if large numbers of IT professionals are so highly trained that they simply migrate to developed countries.[32]

There are two reasons for not stopping here, however. The first is a defensive one: greater automation of software development and the emergence of other low-labour-cost sources of competing IT skills[33] may lead to export growth falling or even reversing as global demand for Indian programming services slows or falls. The second reason is that it may be possible to do even better. Comparative advantage is not fixed, but can evolve over time. There are now numerous formulations of endogenous determination of innovation and growth with international trade, and they suggest that countries may be able to move towards producing higher value-added goods and services as they grow, with favourable consequences for long-run growth.[34] Applying these models is not an automatic proposition, since results are sensitive to assumptions. For example, learning by doing in manufacturing (including software production in this abstract conception) gives different outcomes than the assumption of a separate R&D sector that competes with manufacturing for skilled labour.

If we accept the potential theoretical benefits of moving up the value-added ladder, what does this mean in practice for India's software industry? One possibility is offering higher value-added component services, involving design and strategy. Another is offering more complete packages or bundles of services. The latter differs from the former in that a higher management component is included in the package than in particular aspects of software development, even if those require more technical skill. While the slowdown in the USA that began in 2001 will hamper these developments in India, there is no reason why Indian software firms cannot enter such markets. A growing number of professionals with combinations of engineering and management skills will help in this area.[35]

Companies such as TCS have long-standing domestic and developing-country consulting expertise, but they may be less-suited to compete in a crowded US market than India's new software giants such as Infosys and Wipro, or smaller newcomers such as TechSpan, MindTree and Planetasia.[36] What is the possible competitive advantage of these firms? Certainly lower costs will help, but these may be better used to provide upgraded or broader services rather than in competing on price alone. In particular, empirical research (Banerjee and Duflo, 2000) shows that reputation effects are quite important for Indian software exporters. Such companies may also develop a strong niche in other developing countries, where lower prices may be more important. A high-tech slowdown in the USA may actually aid the development of such firms in India, if it leads to some reversal of the brain drain.[37] Otherwise, managerial and high-level technical skills may constrain movement up the value-added ladder. In some cases, including consulting as well as IT-enabled services, multinational firms have relaxed some of the managerial constraints through their own entry, importing managers as well as training local ones.

The domestic market

The domestic market for IT products and services is certainly not independent of the export market. To the extent that Indian software firms can compete successfully abroad, they can also succeed in their own backyard. In fact, they should have an advantage in the domestic market, in knowing their customers better, and being closer to them. On the other hand, a poor domestic infrastructure, dependence on imported hardware, late-mover disadvantages, and lack of economies of scale and learning by doing, can all reduce or eliminate any advantage that Indian software firms might have over foreign competitors.

Two mitigating factors operate on potential disadvantages of Indian firms. First, some of the problems are faced by all firms, irrespective of location: for example, entering the market for desktop operating systems in the face of Microsoft's dominance is difficult, if not impossible, for any firm anywhere in the world. Second, the boundary lines between domestic and foreign can be blurred when multinationals have Indian subsidiaries, particularly for IT or IT-enabled services. In such cases, the effects on the local economy are not that different from when these services are provided by Indian firms. Two differences in the case of multinationals, however, are in profit repatriation and the creation of another brain-drain channel, if Indian employees of multinationals can be assigned to other countries (as often happens).

At the level of business software and software services, therefore, it seems that issues for the domestic market boil down to the same concerns as for export markets. These are availability of the key inputs, namely various types of skilled IT personnel and managerial and marketing skills. Location and ownership are not of direct importance, but are only proxies for whether the IT software and services provider has the right combination of people, knowledge, experience and reputation to compete successfully. Liberalization in general means that all Indian firms face the challenge of building such combinations of assets. The software industry happens to be significant because it has developed independently of India's traditional business houses, and hence mostly free of the bad habits those business groups developed over many decades of operating in non-competitive environments.

The nature of information goods in general is that they involve high fixed costs of production and low marginal costs. While customization and service provision mitigate this property, they do not negate it. Reputation and experience effects, on the other hand, enhance economies of scale and scope. Hence it is important for Indian software firms to compete simultaneously in domestic and export markets, in order to take advantage of these economies. This is true even though the product – service mix that is being sold in different markets is going to be somewhat different.

We have focused the discussion so far on software. Hardware may offer additional opportunities to Indian IT firms in the domestic market. In developed countries, particularly the USA, the establishment of the PC market took place well before the Internet took off. PC prices are low enough relative to incomes that a large number of households prefer to have an all-purpose device that enables Internet access, rather than a more limited but cheaper 'network appliance'. Internet access (providing access to large quantities of information, entertainment and an inexpensive communications channel) is probably the most attractive use for many potential consumers of IT in India, but Internet penetration may not go far enough with hardware designed for developed countries. Internet cafes with shared access are a partial solution, but their economics may still restrict them to middle-class households.

The possibility of designing and building lower-cost access hardware in India may represent an opportunity for the domestic IT industry. While India has tried to develop a domestic hardware industry since the 1980s, it has not succeeded in establishing an industry that is efficient and globally competitive.[38] Much of the problem has lain with lack of scale and infrastructure. For components such as sophisticated chips this will continue to be the case, but, as noted earlier, Indian industry can build on existing capabilities in assembly of standardized components. If some of these components are designed specifically for the broader Indian market, for example to go into low-cost Internet and telecom access devices, as envisaged by the IIT Chennai group, where they are built may not be crucial.[39] We note once more that Dell is a profitable company because it serves targeted markets efficiently, not because it manufactures sophisticated components. Instead, management and infrastructure are the key inputs that are required.

Broad-based growth

Are a software industry that serves the domestic market as well as exporting, a hardware industry that can produce low-cost access devices, and IT-enabled services for foreign markets, together enough for broad-based economic growth? Clearly the IT industry alone can only contribute one or two percentage points to India's growth rate. The concern is that it will remain an enclave, exacerbating inequality, and doing little for long-run growth. Even in the USA, tall claims for the benefits of IT for the broader economy, through enhanced productivity growth, are met with skepticism, and the evidence from varied studies is mixed.[40]

While one may hesitate, therefore, to claim that IT is the answer to India's ills, there are reasons to argue that it potentially has beneficial effects beyond its direct contribution to growth. The argument one can make is rooted in standard analysis from development economics.[41] To the extent that IT can have significant effects on the efficiency of operations in other industries, there are strong *complementarities* between the IT sector and the rest of the economy. Examples of areas where increased efficiency may be possible include accounting, procurement, inventory management and production operations.[42] This is, of course, the standard argument in the USA for the virtues of the 'new economy' based on the Internet. It should be recognized that the difference for India is that it is starting from a much lower level of IT-adoption, and the potential gains may be higher. In fact, developing countries have the opportunity to leapfrog over older, more expensive approaches such as Electronic Data Interchange, which represent significant legacy investments in countries such as the USA.

The usual concern with IT-adoption is job loss, and there is certainly the potential that certain kinds of clerical jobs will be eliminated or reduced in numbers. Unions in Indian industries such as banking have opposed 'computerization' for this reason. However, the evidence suggests that increases in other kinds of jobs as a result of IT-use more than make up for job loss, so that total employment is not a significant issue. This leaves the issue of adjustment costs, and here severance pay rules or government job adjustment assistance can be more effective and efficient than the current morass of detailed restrictions embodied in India's labour laws. This is why Desai (2000) is right to stress the importance of broader labour law reform if the benefits of growth in India's IT sector are to be fully realized.

In the context of complementarities, it is also important to recognize that these effects are not just in terms of cost savings. IT implementation may enhance the quality of service beyond anything that is feasible through other methods. Furthermore, depending on who the 'customers' are, the benefits may accrue to a broad cross-section of the population. Improved efficiency in the stockmarket as a result of automated trading and settlement may benefit a small section of the population (though the indirect benefits of greater capital market efficiency may be broader). The use of IT in banking may impact only the middle classes.[43] However, the computerization of the Indian Railways' reservation system has had tremendous benefits for the masses who use this mode of transportation. To summarize, the existence of these benefits, and of efficiency gains, together representing the working of complementarities or linkages, is an argument against the 'enclave' view of the IT sector in India.

Note that potential benefits do not necessarily translate into actual ones; firms and managers can make mistakes. However, this is no different from any other kind of investment, and the point is that in a reasonably competitive industry with sufficient information available, there is always pressure to make the right decisions, rewards for those who do, and punishments for those who do not. Indian industry must be allowed to follow this model to realize the potential benefits of IT. If it is discouraged from making such investments, the domestic market for Indian IT will not grow, with negative consequences for the IT sector as a whole.

We have focused implicitly so far on the formal, organized or 'large-scale' sector in assessing the impact of IT. Even if the growth of the IT has positive spillovers for other industries, this leaves out a substantial portion of the economy. We postpone a discussion of government use of IT to a separate section, but now turn to issues of truly broad-based impacts of IT. There are two related, but separate areas of impact. First, information processing may enhance efficiency in agriculture as well as in manufacturing. While individual farmers cannot make IT investments, agricultural cooperatives can provide the institutional framework that allows farmers to benefit. For example, Chakravarty (2000) gives the example of IT use at milk-collection centres in cooperative dairies. This permits faster and safer testing, better quality control, quicker and more accurate payments to farmers, and time savings for farmers in their deliveries. The falling cost of information processing means that such success stories can potentially be widely replicated.

The second impact is in the communication of information. Here the case studies are legion. Farmers and fishermen can receive weather forecasts, market price quotes, advice on farming practices, and specific training. Offers to buy or sell livestock, or other two-way communications, are also possible. Some of this information dissemination and exchange is best done through voice media, while other types require the capabilities of the Internet. Some evidence suggests, not surprisingly, that richer farmers and fishermen, as well as middlemen, are faster adopters of such technologies (*The Economist*, 2001a), but falling access costs, through innovations such as those of the IIT Chennai group, should broaden information access and its benefits. Broad-based benefits of IT require broad-based access to the network, as we have already discussed.

Ultimately, the central feature of the information revolution has been the dramatic fall in the costs of information processing, storage and transmission (see Table 2.2). It is this fundamental technological fact that makes IT special or different from other sectors. It is this fact that makes IT of potential value to the masses of a poor country such as India. The challenge is to translate this potential into cost-effective, valued products and services. Meeting this challenge requires a policy environment within which entrepreneurial energy can flourish.

Economic policies

The issue of economic policy is obviously much larger than we can adequately tackle here. However, we will briefly comment on general

Table 2.2 Falling costs of computing ($)

Costs of computing	1970	1999
1 Mhz of processing power	7 601	0.17
1 megabit of storage	5 257	0.17
1 trillion bits sent	150 000	0.12

Source: Pam Woodall, 'The New Economy: Survey', *The Economist*, 23 September 2000, p. 6, chart 1.

microeconomic and macroeconomic policy issues, and then address specific policies for the IT sectory. The two areas may, of course, overlap. For example, the growth of the IT sector and of software exports in particular, may make certain kinds of general policy questions more salient.

We noted earlier (p. 27) the central areas of India's policy reforms: replacing quantitative trade restrictions with tariffs, lowering effective levels of protection, removing an area of discretionary controls on private-sector investment, and the creation of modern financial markets. Standard examples of where these reforms can be built upon to further stimulate growth include removal or relaxation of obsolete 'small-scale sector' reservations and size restrictions, privatization of inefficient state-owned enterprises, rationalization of tax-subsidy policies and tax administration, and relaxation of severe labour-market restrictions. This list can be characterized by its emphasis on improving the efficiency of the mechanisms with which the government directly affects the private sector. The entire Indian economy, not just the IT sector, can presumably benefit from such reforms which will reduce distortions of private-sector behaviour.

A second area where attention is required may be characterized as enabling reforms. These include reforms of contract law and judicial institutions; financial sector regulatory institutions; telecom sector regulatory institutions; infrastructure such as electric power, roads and ports; and systems of education and training in general.[44] Again, the benefits of such reforms are potentially quite general, and not restricted to any one sector of the economy.

A third area of policy is macroeconomic management. While India's record here is quite good, it needs to make a transition in its policy institutions here as well, since removing detailed microeconomic controls requires changes in the regulatory modes of macroeconomic management. Perhaps the area that has received the most attention is policies towards international capital flows and their implications for exchange-rate management. Desai (2000) has suggested that large projected increases in software exports could create a 'Dutch disease' phenomenon,[45] in which a resulting exchange-rate appreciation hurts other sectors, and revenues from exports are wastefully spent. Several factors mitigate this concern: the likelihood that export revenue growth will slow down from recent levels; the substantial linkages that exist between software, the IT sector as a whole, and the broader economy (unlike natural-resource extraction enclaves); and a better understanding of exchange-rate management than existed 25 years ago, when the phenomenon was first identified and labelled. Our conclusion, therefore, is that while exchange rate policy is certainly important in general, the growth of the IT sector will not necessarily raise special concerns.

Given that there is plenty that remains to be done in terms of overall economic policy reform, are there areas where the IT sector deserves special attention? The answer we give here, with one partial exception, is 'no'. Special subsidies or export incentives are likely to be inefficient ways of stimulating the growth of the IT sector, or of positive spillovers for the rest of the economy. Similarly, special central government initiatives to increase the availability of IT training and related education are also likely to represent a mistargeting of scarce government resources. The same stricture applies, to some extent, to state government policies to encourage the IT sector.[46] In these cases, the government may be better off removing general restrictions to doing business, as well as providing an enabling institutional infrastructure (appropriate laws and regulations), rather than attempting to target the IT sector through a form of industrial policy. Thus, for example, allowing greater private (particularly IT industry) initiative in training and certification of IT professionals makes more sense than the creation of new government institutes.

The partial exception lies in the telecoms sector, which, as discussed earlier, has particularly strong complementarities with the broader IT sector. Development of the telecoms sector is certainly critical, but rather than setting unrealistic quantitative goals and trying to bureaucratically manage change, policies to achieve development goals would do better to emphasize removing barriers to innovations that will support lower-cost access to telecom networks of all kinds (wireless and fixed, voice and data). Interconnection charges and licensing fees are the main policy instruments that need attention in this respect, encouraging competition rather than forcing unrealistic cross-subsidization targets on large private monopoly or duopoly licensees.

Governance

One area where government can provide indirect support for the IT sector is by boosting the domestic market though its own purchases. Of course purchases of sophisticated equipment and software that sits unused in high-level bureaucrats' offices will have little positive impact. However, there are reasons to be more optimistic about the use of IT in government.

There are two broad uses of IT for improved government functioning. First, back-office procedures can be made more efficient so that internal record-keeping, flows of information and tracking of decisions and performance can be improved. Second, when some basic information is stored in digital form, it provides the opportunity for easier access to that information by citizens. The simplest examples would be e-mailing requests or complaints, checking regulations on a web page, or printing out forms from the web so that a trip to pick up the forms from a physical office can be avoided. More complicated possibilities are checking actual records, such as land ownership or transactions. Still more complicated are cases where information is submitted electronically by the citizen.

What is encouraging is the number of examples of successful pilot e-governance programmes that have made some of the above actions possible. These examples include:

- Computer-aided registration of land deeds and stamp duties in Andhra Pradesh, reducing reliance on brokers and possibilities for corruption.
- Computerization of rural local government offices in Andhra Pradesh for delivery of statutory certificates of identity and landholdings, substantially reducing delays.[47]
- Computerized checkpoints for local entry taxes in Gujarat, with data automatically sent to a central database, reducing opportunities for local corruption.
- Consolidated bill payment sites in Kerala, allowing citizens to pay bills under 17 different categories in one place, from electricity to university fees.
- E-mail requests for repairs to basic rural infrastructure such as hand pumps, reducing reliance on erratic visits of government functionaries.[48]

While the extent of such programmes, and therefore their benefits, are still very limited, one of the essential characteristics of information is that it is a non-rival good. Hence there is every reason to believe that the economics of such projects will continue to improve, particularly if costs of access continue to fall. It is important to note that once Internet access is available, its benefits are not restricted to e-governance. Individuals can obtain market information, training, job information, advice on farming techniques, and so on, as discussed earlier in this section. There are economies of scope as well as of scale. Also, Metcalfe's 'Law', that the value of a network is approximately proportional to the square of the number of users, probably holds with full force in many Indian contexts.[49] Denser local communications networks are likely to aid economic activity in general.

Given the poor quality of governance in India, one is therefore inclined to strongly favour e-governance initiatives that provide direct benefits to citizens, particularly those who are less well off (the rich in any case hire intermediaries to collect information, make payments, etc.). Here we suggest that in addition to the direct benefits, there may be advantages through support of the growth of the domestic market. Increasing the size of the market while encouraging competition may be more cost-effective for the government than any direct subsidies or incentives to the IT industry.

Conclusion

Our goal in this chapter has been to assess the possible role of India's IT industry as a driver of higher economic growth in India, without exacerbation of inequalities or creation of instability. Our conclusion is cautiously positive. While projections for software exports may be over-optimistic, complementarities or spillovers in the domestic market, including increased government use of IT, are likely to be strong. For this rosy scenario to play out, however, continued broad economic reforms will be important, as well as reforms in the telecoms sector that promote competition and innovation in providing last-mile access. These reforms include reductions of licensing barriers to entry and high interconnection charges for higher levels of the network. Such enabling reforms are more likely to support broad-based, sustainable growth than narrowly targeted incentive schemes.

Notes

1 Alternative terms include ICT, for information and communications technology, and ICE, for information, communications and entertainment. I am grateful to Atanu Dey for these suggestions.

2 Rafiq Dossani has pointed out to me (personal communication) that the distinction between hardware and software has blurred, as some 'hardware'

firms have focused on design and outsourced their manufacturing. Design of hardware involves the writing of software code, and this designing ability can be used for pure software tasks. Our subsequent discussion includes the distinction between design and manufacturing (including assembly).

3 The data are from NASSCOM (2000).

4 This is based on using present payments on current account, and projecting them to grow at the present rate of 7.6 per cent per annum, as suggested and calculated by Desai (2000).

5 See, for example, Heeks (1996, 1998) and Desai (2000).

6 The figures are from *Dataquest* (2001). Similar figures are given by Heeks (1998) and Desai (2000).

7 See *DQ Week* (2001).

8 See *The Economist* (2001).

9 See *Business Week* (2001). However, Aggarwal (2001) gives a substantially lower figure of 55 000 engineering graduates annually, excluding private institutes and Masters of Computer Applications (MCA). Arora and Arunachalam (2000) estimate an overall figure, including MCAs and graduates of informal training institutes, of close to 140 000 per annum. See also Arora *et al.* (2001a, 2001b) and Saxenian (2001).

10 See Arora and Arunachalam (2000), as well as Heeks (1996) and Desai (2000) for further discussion. Desai provides a detailed discussion of manpower requirements by type of task.

11 See Aggarwal (2001).

12 Arora *et al.* (2001a) construct a lower estimate of $15 600 for 1998–99 (their Table 1). However, revenues doubled over the next two years, implying a figure closer to our estimate.

13 The implication is that changes in the product–service mix towards that involving higher value-added tasks would be associated with these improvements, resulting in increased productivity.

14 Rafiq Dossani has expressed a slightly different view (private communication). He argues that there are a range of IT skills that are needed, and the IITs are not necessarily the best sources for lower-end skills. On the other hand, all kinds of new training and certification institutes may be good alternatives. Our point here applies more narrowly, to the best use of very limited government funds.

15 The issue here goes beyond the temporary H1-B visa boom in the USA, now reversed. The best from the IITs have been going to the USA for higher education for decades, and only a small fraction return. While part of the problem is simply the extreme subsidy given to higher education in India, other policy factors play a role, and the example of Taiwan shows that the brain drain can be arrested (Saxenian, 2000).

16 Again, one must acknowledge the problems of electric power and transportation also. To some extent, software avoids problems of physical transportation, but power remains a critical input.

17 See Heeks (1996, 1998).

18 Only 12 VPTs were installed in the required time period.

19 See Telecom Regulatory Authority of India (2000). Yale Braunstein has pointed out to me that this wide spread partly reflects an inefficient tariff structure, with domestic long distance and international calling being priced significantly above cost.

20 An anecdotal example from my field research in Bathinda district of Punjab in December 2001 can illustrate: a farmer told us he had taken computer lessons, bought a home computer, and signed an Internet service contract so that he could exchange e-mail with his brother in Toronto, Canada. All three IT-related products and services depended on basic telecom availability. See also Prahalad and Hart (2002).

21 See Jhunjhunwala (2000), as well as numerous presentations available on his web site, at www.tenet.res.in/ashok.html. The overall group is called the Telecommunications and Computer Network (TeNeT).

22 The innovations are described in more detail in Jhunjhunwala (2000). They include some wireless components, combined access to voice and Internet connections, and low-cost access devices.

23 This does not have to mean trying to do it all in-house. India will probably never have a comparative advantage in the large-scale production of various kinds of chips, but efficient design and assembly of devices that use these chips is certainly feasible: again Dell is a useful example to bear in mind.

24 In particular, it may be efficient to sacrifice revenue for lower-cost provision, as in the case of Hong Kong's approach: I am grateful to Rafiq Dossani for this example.

25 According to Ashok Desai (personal communication), the 35 telecom circles that made up the franchise areas do not generate enough traffic for current economic sustainability, with the exception of four metro circles. Revenue from inter-circle interconnections goes entirely to the DoT. However, the economics of the situation may change if set-up costs are brought down substantially, as innovations discussed below may achieve.

26 The states in India are comparable in population size and area to European countries, though of course not in terms of income.

27 Atanu Dey (personal communication) has pointed out to me that the solution that many countries seem to be converging on in the case of interconnection is that of 'light-handed regulation'. The parties involved privately negotiate their interconnection terms, and only in case of impasse does the regulator step in.

28 I am grateful to P.D. Kaushik (personal communication) for this point – by his count, there are over 400 central government statutes governing manufacturing, as well as numerous state laws. He also notes that the software industry escaped these constraints partly by not being recognized by the government as an industry by itself.

29 See Dossani and Manikutty (2000) for details.

30 See for example, the paper by M.S. Verma, Chairperson of the TRAI (Verma, 2000).

31 Note that, to the extent that India is providing intermediate goods or services in its software exports, the situation is more complex than that of standard trade theory where only final goods are traded.

32 As we noted earlier, Desai (2000) is the source of these perceptive observations.

33 These include other Asian countries such as China and the Philippines, but also the countries of the former USSR.

34 In particular, see Grossman and Helpman (1991) and Rivera-Batiz and Romer (1991a, b).

35 Similar considerations apply to IT-enabled services, where there is also a value-added ladder going from data entry and conversion through rule-set

processing, problem-solving and direct customer interaction, to expert 'knowledge services'. See *The Economist* (2001b), p. 60, where this classification is credited to Raman Roy.

36 See Das (2001) for a discussion of the Indian e-business consulting market.

37 For a description of how Infosys is struggling with the tech slowdown, to 'transform itself... from a great code-writer to a one-stop technology services provider', see Bjorhus (2002).

38 See Hanna (1994), for example, for further details.

39 The most immediate example of this possibility is the 'Simputer', designed by scientists from the Indian Institute of Science. The portable $200 device will use some parts manufactured in Singapore, Linux-based software developed in India, and run on 3 AAA batteries.

40 See, for example, Pohjola (1998) and Woodall (2000).

41 A general discussion of the concepts presented next can be found in Ray (1998), which also provides detailed source references.

42 These are all examples of what are also known as 'forward linkages', since IT adoption has positive impacts on the operations of a range of industries. The effect of the growth of the IT sector on the provision of technical education would be an example of a 'backward linkage'. In either case, there is a complementarity at work.

43 P.D. Kaushik has suggested to me that rural banking and other financial services may be substantially improved through the use of IT.

44 Problems of primary education have been well-recognized: see Dreze and Gazdar (1997) for example. While higher education has been *relatively* favoured, the public monopoly here is also failing to deliver effectively, as pointed out perceptively in a recent piece by Panagariya (2001).

45 The best simple explanation of Dutch disease I have come across is by John McLaren, at www.columbia.edu/~jem18/teaching/pepm/dutchdis.pdf. McLaren clarifies the source of concerns that are associated with Dutch disease, including exacerbation of *prior* distortions, and of inequality.

46 Bangalore in Karnataka is well-known as a regional IT centre in India, having developed initially without much explicit government support. The governments of Andhra Pradesh (Eischen, 2000) and Tamil Nadu (Bajpai and Radjou, 1999, and Bajpai and Dokeniya, 1999) have led in attempts to establish IT-based industries with conscious government policies. Other state governments, such as Punjab (see www.dqindia.com/mar1599/news.htm) are following suit.

47 These two examples are from Bhatnagar and Schware (2000), which also provides broader examples, including ones driven more by the efforts of NGOs than governments.

48 These three examples are from *India Today* (2000), which also lists several other similar projects.

49 Ashok Desai raised the issue of how Metcalfe's 'Law' and economies of scale can be consistent with a model of small private competitors. I think that a distinction that helps here is between ownership and usage. Usage networks are where demand-side economies of scale matter. On the other hand, on the supply side, ownership networks can be smaller as long as interconnection charges are regulated appropriately.

References

Aggarwal, B.B. (2001) 'Faculty Scarcity at IITs Threatens Knowledge Capital', 19 March, http://www.ciol.com/content/news/trends/10103902.asp

Arora, A. and Arunachalam, V.S. (2000) 'The Globalization of Software: The Case of the Indian Software Industry'. A report submitted to the Sloan Foundation. Carnegie Mellon University, Pittsburgh PA. http://www.heinz.cmu.edu/project/india/publications.html

Arora, A., Arunachalam, V.S., Asundi, J. and Fernandes, R. (2001a) 'The Indian Software Service Industry', *Research Policy*, vol. 30, pp. 1267–87.

Arora, A., Gambardella, A. and Torrisi, S. (2001b) 'In the footsteps of the Silicon Valley? Indian and Irish Software in the International Division of Labour', Paper presented at the conference, 'Silicon Valley and its Imitators', Stanford Institute for Economic Policy Research, July 2000.

Bajpai, N. and Radjou, N. (1999) 'Raising the Global Competitiveness of Tamil Nadu's Information Technology Industry', Development Discussion Paper no. 728, October, Harvard Institute for International Development.

Bajpai, N. and Dokeniya, A. (1999) 'Information Technology-Led Growth Policies: A Case Study of Tamil Nadu', Development Discussion Paper no. 729, October, Harvard Institute for International Development.

Banerjee, A.V. and Duflo, E. (2000) 'Reputation Effects and the Limits of Contracting: A Study of the Indian Software Industry', *Quarterly Journal of Economics*, vol. 15(3), pp. 989–1017.

Bhatnagar, S. and Schware, R. (2000) *Information and Communication Technology in Development: Cases from India*. New Delhi: Sage.

Bjorhus, J. (2002) 'India's Infosys Struggles through Transformation', *San Jose Mercury News*, 11 February, p. 1E.

Business Week (2001) 'India 3.0: Its Software Outfits Take On the World', 26 February, pp. 44–6.

Das, S. (2001) 'The Indian Challenge: Will They ... Or Won't They?', 7 March, http://voicendata.com/content/top_stories/101030703.asp

Dataquest (2001) 'SW INDUSTRY: Working Around the Slowdown', 14 February, http://dqweek.ciol.com/content/search/showarticle.asp?artid=21244

DQ Week (2001) 'Medical Transcription: Not in the Pink of Health', 2 February, http://www.ciol.com/content/search/ showarticle.asp?artid=21128

Desai, A.V. (2000) 'The Peril and the Promise: Broader implications of the Indian Presence in Information Technologies'. Working Paper, August, CREDPR, Stanford University.

Dey, A. (2000) 'New Telecom Policy 1999: A Critical Evaluation', Paper presented at Conference on Telecommunications Reform in India, Asia/Pacific Research Centre, Stanford University, 9 & 10 November.

Dossani, R. and Manikutty, S. (2000) 'Reforms in the Telecommunications Sector in India: An Institutional View'. Paper presented at Conference on Telecommunications Reform in India, Asia/Pacific Research Centre, Stanford University, 9 & 10 November.

Drèze, J. and Gazdar, H. (1997) 'Uttar Pradesh: The Burden of Inertia', in A. Sen and J. Drèze, *Indian Development: Selected Regional Perspectives*. Delhi: Oxford University Press.

Economist, The (2001a) 'Another Kind of Net Work: Mobile Phones in India', 3 March, p. 59.

Economist, The (2001b) 'Outsourcing to India: Back Office to the World' 5 May, p. 59.

Eischen, K. (2000) 'National Legacies, Software Technology Clusters and Institutional Innovation: The Dichotomy of Regional Development in Andhra Pradesh, India', University of California, Department of Sociology.

Grossman, G. and Helpman, E. (1991) *Innovation and Growth in the Global Economy*. Cambridge, Mass.: MIT Press.

Hanna, N. (1994) 'Exploiting Information Technology for Development: A Case Study of India'. World Bank Discussion Paper no. 246.

Heeks, R. (1998) 'The Uneven Profile of India's Software Exports'. IDPM Working Paper no. 3, October, University of Manchester.

Heeks, R. (1996) *India's Software Industry: State Policy, Liberalisation and Industrial Development*. New Delhi, Thousand Oaks and London: Sage.

India Today (2000) 'Is e-Governance for Real?', 11 December, pp. 70–5.

Jhunjhunwala, A. (2000) 'Unleashing Telecom and Internet in India'. Paper presented at the Conference on Telecommunications Reform in India, Asia/Pacific Research centre, Stanford University, 9 & 10 November.

National Association of Software and Service Companies (2000a) 'Indian IT industry. New Delhi', http://www.nasscom.org/template/indianitss.htm

National Association of Software and Service Companies (2000b) 'Indian Software and Services Industry. New Delhi', http://www.nasscom.org/template/itinindia.htm

Panagariya, A. (2001) 'Abolish this Monopoly', *Economic Times*, 28 March.

Pohjola, M. (1998) 'Information Technology and Economic Development: An Introduction to the Research Issues', WIDER Working Paper no. 153, November, United Nations University.

Prahalad, C.K. and Hart, S.L. (2002) 'The Fortune at the Bottom of the Pyramid', http://www.strategy-business.com/media/pdf/02106.pdf

Ray, D. (1998) *Development Economics*. Princeton: Princeton University Press.

Rivera-Batiz, L.A. and Romer, P.M. (1991) 'Economic Integration and Endogenous Growth', *Quarterly Journal of Economics*, vol. 106(2), pp. 531–55.

Rivera-Batiz, L.A. and Romer, P.M. (1991) 'International Trade with Endogenous Technological Change', *European Economic Review*, vol. 35(4), pp. 971–1004.

Saxenian, A. (2000) 'The Bangalore Boom: From Brain Drain to Brain Circulation?', in K. Keniston and D. Kumar (eds), *Bridging the Digital Divide: Lessons from India*. Bangalore: National Institute of Advanced Study.

Saxenian, A. (2001) 'Bangalore: The Silicon Valley of Asia?', Working paper no. 91, Centre for Research on Economic Development and Policy Reform, Stanford University, February.

SGKI (2000) Subject Group on Knowledge-based Industries, Prime Minister's Council on Trade and Industry. *Recommendations of the Task Force on Knowledge-based Industries*. Prime Minister's Office, New Delhi. http://www.nic.in/pmcouncils/reports/knowl/

Telecom Regulatory Authority of India (2000) 'Consultation Paper on Issues Relating to Universal Service Obligations'. TRAI, New Delhi, 3 July.

Verma, M.S. (2000) 'TRAI's Objectives and Policy Focus in a Changing Environment'. Paper presented at the Conference on Telecommunications Reform in India, Asia/Pacific Research Centre, Stanford University, 9 & 10 November.

Woodall, P. (2000) 'The New Economy', *The Economist*, 23 September, Survey p. 6.

3

Facing Hypercompetition in World Software Markets: Global Strategies for India

Jati K. Sengupta

Introduction

The microcomputer industry today has reached a new phase: the phase of hypercompetition. Hypercompetition adds three dynamic facets to the standard model of market competition. The first is the *innovation efficiency*. Innovation is very broadly defined as in Schumpeterian dynamics for example, it may refer to a new product, a new process or technology, or opening up a new market niche. By this efficiency the firms generate new knowledge and new applications. Old knowledge gets replaced, new ones displace the incumbents and this competitive pressure grows continually. The second facet is *access efficiency*, whereby firms race up the escalation ladder in the strongholds arena. Today the Internet economy with its e-trade, e-commerce and dramatic changes in transportation and communications has transformed the demand market globally. Goods and services made by a company become available anywhere in the world today, and customers have access to a wide variety of goods and sellers. Networking and mergers provide firms of today with immense opportunities to augment their strongholds arena, which may help them counteract the threat of new entrants. Aggressive competition to improve access efficiency thus constitutes the core of a successful firm's winning strategy. Finally, the key driver of today's hypercompetitive shift in market competition is the dynamic resourcefulness of an industry or the ease with which new strategic assets can be created and developed. This may be called *dynamic resource efficiency*. Note that this process of creation of new strategic assets at various points of the value chain is driven by rivalry among a shifting set of competing firms. The key point is to understand that rivalry in this setup is very productive and not at all harmful, since it creates pressure on companies to innovate and improve. Local rivals push each other to lower costs, improve quality and service, and create new products and better services. Also, demanding buyers pressure companies to meet higher standards in goods and service. Note that this dynamic resource efficiency poses a sharp contrast to the static world of competition, where rivalry is most harmful to static resource efficiency. Thus static competition or rivalry in an industry works continually to drive down the risk of return and threaten profits. This stifles any further growth of the industry. In hypercompetition, firms have to realize that rivalry entails a positive-sum game, which helps the long run-growth of a successful firm.

The software industry is closely dependent on the microcomputer industry, and the rapid growth of the latter in recent decades has accelerated the demand for the products and services of the software industry, which is also very hypercompetitive. Within the overall set of information technology (IT) and microcomputer or personal computer (PC) industry in particular, software production is a vital input, embodied in a vast and highly diversified range of products and services. It has positive externality effects on other sectors and it is highly skill-intensive. It has played a key role in the rapid growth of the so-called NICs (newly industrializing countries) of SouthEast Asia, which have grown steadily at the rate of 5 per cent or higher per year over the last two decades. These features of the software industry (SI) make it uniquely suitable for expanding and developing its role in the Indian economy today. However, due to the hypercompetitive nature of the PC and SI markets in the world today, there is an urgent need to develop new types of strategies that are most suitable for the hypercompetitive age.

Our object here is to discuss the salient features of the hypercompetitive market structure in IT-based industries in the USA today and to propose a set of production and marketing strategies for a developing country like India. The proposed strategies are broad-based for the whole economy and not specifically oriented to corporate policies. This proposal assumes two preconditions. One is the outward orientation of India's international trade as with NICs of SouthEast Asia. The other is the pursuit of a steady and sustained drive for industrial growth as a common goal of the private and public sectors in India. These two preconditions are essential for this new competitive framework in the world economy today.

New paradigms

The new economic order emerging today, sometimes called the *new economy*, is spreading all over the world. This is nothing short of an industrial revolution, and it is a revolution in information explosion and in knowledge capital. Three key elements of this revolution are: increasing efficiency of the microcomputer industry; intra-firm and inter-industry diffusion of knowledge and new innovations in the Schumpeterian sense; and the global expansion of trade through network externalities. It is of some importance to analyse the impact of these key elements on the rapid and steady development of the US economy over the last decade. According to the recent estimate by Jorgenson and Stiroh (2000), the price decline for computers has accelerated in recent times, reaching nearly 28 per cent per year from 1995 to 1998. Exploitation of significant scale economies has been the key to this rapid price decline. In response, investment in computers has exploded in the USA today and the overall national growth contribution of computer hardware has increased more than fivefold, to 0.46 percentage points per year in the late 1990s. Software and communications equipment contributed an additional 0.30 percentage points, and more recent estimates until 2000 reveal further increases.

As the production of computers improves and becomes more efficient, more computing power is being produced from the same inputs that is, learning by doing. This increases the overall productivity in the computer-producing industry and contributes to growth in total factor productivity (also called technological progress) for the economy as a whole. Labour and R&D productivity also grow at a faster rate. Secondly, the computer-using industries are now using skilled labour working with more and better computer equipment and this investment increases labour productivity in these industries further. Thus the computer industry has experienced a sustained increase in returns to scale (RTS) and in technological progress. In their empirical estimate, Norsworthy and Jang (1992) found for the US computer industry as a whole, the level of RTS as 1.400 for the period 1973–80 and even higher in recent years. The productivity growth (technological progress) for the electronic equipment industry (which includes the computer industry) over the period 1958–96 was 2.0 per cent per year on average. For the more recent period 1992–98 the comparable figure for the computer industry (mainframe and microcomputer) exceeded 2.5 per cent per year. The major source of productivity growth has been the growth in demand over the last two decades, which has exceeded 12 per cent per year on average.

The rapid growth episodes of NICs in SouthEast Asia exhibit the impact of openness in international trade and demand growth on the high growth rate of these countries. For example, the average annual growth rate of real national income in Korea over the period 1970–73 was 16 per cent, of which export expansion contributed 55.7 per cent and domestic demand expansion 51.9 per cent. By comparison Mexico contributed about 5.9 per cent in export expansion to the annual growth rate of 6.1 per cent. India's contribution is much lower than that of Mexico.

Recently, Kraemer and Dedrick (1994) analysed the pattern of investment in information technology (IT) industries of 11 countries of the Asia-Pacific region over the years 1984–90, where the spending on computer hardware and software services was used as a measure of IT investment. They found the average growth in IT investment over 1984–90 was very high for Korea (24.49 per cent), Taiwan (21.64 per cent), Singapore (18.06 per cent) and Hong Kong (15.22 per cent), whereas for slower growth countries it has been much lower, for example Malaysia (10.77 per cent) or Philippines (12.21 per cent). The infrastructure investment in Korea, Taiwan and Singapore has considerably helped the growth of IT investment in the higher performing countries of Asia. As Dahlman and Nelson (1995) have shown, the four East Asian high performers Singapore, Taiwan, Hong Kong and Korea have consistently used the following strategies to capture the international diffusion of technical know-how – extensive use of technology licensing and links with foreign subsidiary firms; and large-scale contact with foreign buyers and the development of strong high-tech industries embodying the latest technical know-how from the USA. Taiwan's growth over 1985–2000 followed a much different pattern than Korea or Japan. While Japan and Korea followed a policy of increasing the average size of firms and thus increasing the degree of industrial concentration, Taiwan increased the number of firms or establishments in the manufacturing sector, including the software and microelectronic industries.

The growth of IT investment in the USA has been most rapid over the last two decades, exceeding on average 15 per cent per year. It has outpaced the growth of demand, and as a result the average rate of return on all assets for the computer industry as a whole increased from an annual rate of 31.1 per cent to 34.5 per cent over the period 1975–84 according to Norsworthy and Jang's estimate. For the more recent years 1985–98, it increased still higher. The diffusion of the impact of investment in the computer industry to other sectors such as bioengineering, telecommunications and consumer electronics has been most spectacular in the current decade 1990–2000. The basic trend of this impact has been an all-round increase in industrial productivity, leading to a growth rate in the US economy of 4 per cent or higher per year over the last decade. The new paradigm fueling this rapid growth process can be summed up by three letters – IDI. The first, I, refers to the stream of new innovations in the computer industry, including the software development sector. The second letter, D, for demand, reflects the impact of new innovation on the increase of demand and national income, generating a further round of innovations. The letter, I, for innovations, is used here in a broad Schumpeterian sense – it includes new processes, new forms of knowledge-creation and application and new products such as the development of new and industry-specific software. All IT investments are for innovative goods and services and it augments the demand flow in three ways: one is B to B trade (business to business transactions), the others are domestic and international trade. Shapiro and Varian (1999) have emphasized this point very strongly in the increased demand flow in today's Internet economy fueled by the personal-computer network. They characterized it as the economies of scale in demand. This means that any new innovative investment in the computer industry generates a multiplier effect on demand, both specific and more general, and this growth in demand affects the next flow of innovations through new capacity creation, replacement of old technology and utilization of scale economies. Thus the cycle goes on.

Note that the IDI process is a mere restatement of Adam Smith's thesis that the division of labour is limited by the size of the market. For Adam Smith this is an evolutionary and growth perspective. Dividing the production process into increasingly simpler elements is a *continuous discovery process*, yielding new knowledge with both internal and external effects. Current innovations in the microelectronic industry has expanded the market size in all computer-using industries and this trend is unlikely to slow down anytime soon. This is because the continuous discovery process emphasized by Adam Smith is going on in all other traditional fields and industries. This framework has often been called a state of hypercompetition, which has two aspects: static and dynamic. The static aspect stresses the destructive aspect in the sense that only the fittest survive, that is only the cost-efficient firms remain, others exit. This occurs in a world of fixed and limited demand, where profits are driven to their lowest levels. The dynamic aspect of hypercompetition assumes interactive and increasing demand due to market expansion, and here competition between firms improves the innovative efficiency of all firms. This has sometimes been called

Metcalfe's Law, which states that the value of a network or process goes up as the square of the number of users or suppliers. This implies that a tenfold increase in initial demand leads to a hundredfold increase in the value of a network. Thus the dynamic aspect of hypercompetition is the creative aspect, since it enhances the productive efficiency of the industry as a whole. These two aspects have been analysed by Sengupta (2001a,b) separately in respect of the US computer industry over a 15 year period 1984–99. This analysis has shown the most striking rates of growth of firms such as Intel, Microsoft and Cisco.

Three Cs

What makes a firm grow? What causes an industry to rise? What causes a country to prosper? From a very broad standpoint, two types of answers have been given to these questions. One is managerial, the other economic. The managerial perspective is based on organization theory, which focuses on the core competence or dynamic capabilities as the primary sources of growth of firms, industries and organizations. The economic perspective emphasizes national productivity and efficiency as the basic source of growth. Adam Smith's inquiry into the causes of wealth of nations identified the market or demand as the prime mover of productivity increase across nations. Economic efficiency of both physical and human capital and innovations through improvements in skill and technologies have been considered as the key source of economic growth by the modern theory of endogenous growth, also called the new growth theory.

These three major sources of growth start at the microeconomic level of a firm (enterprise) or organization and spread over the industries across the national economy. These growth sources serve to emphasize the following three principles: the three Cs:

1 Principle of competitive advantage
2 Principle of comparative advantage
3 Principle of core competence

Each of the three principles characterizes a broad framework of strategies for a firm, an industry or the whole economy. Together they form a growth synergy: a cumulative process of sustained growth through a complementary set of principles of action.

The heart of the competitiveness principle is how to improve national productivity. Sustained growth in productivity requires that a nation's firms continually improve productivity in existing industries in several different ways; for example by raising product quality, improving technology, reducing costs and prices and increasing production and access efficiency. Recently, Porter (1990) has emphasized the point that in today's global market competitive advantage may emerge in many forms. For example, firms may gain competitive advantage from conceiving new ways to conduct activities, developing new procedures and technologies and improving productivity levels. Thus Makita in Japan emerged as a leading competitor in power tools because it was the first to develop standardized models in a single plant, that it then sold worldwide. Samsung in Korea rose to prominence in the consumer electronics market in the USA by pioneering new products and clones at cost-efficient prices. Volvo of Sweden maintained a steady share of the highly competitive automobile market in the USA by sustaining product quality, product safety and improving the servicing standards.

Competitive advantage

The principle of competitive advantage emphasized the network of opportunities that exist in a given environment and the dynamic strategies that may be adopted to exploit them. This is essentially the Schumpeterian notion of innovations, which continually shift the strength of competitive advantages from one firm or industry to another, or from one nation to another. Porter (1990) identified four major sources of competitive advantage as innovations which firms can employ to develop viable and successful business strategies in today's global market: these apply specifically to software markets.

1 Moving early to exploit structural change in the market. Early movers gain advantages by reaping substantial economies of scale, reducing costs through learning-curve effects and establishing reliable customer channels through brand loyalty.
2 Perceiving and pursuing innovation. Information technology (IT) today has reached a phase where companies which innovate are frequently not established leaders or even large firms. Even when the innovators are large firms, they are often invaded by new entrants. Software development and related R&D activities provide classic examples. Today's Silicon Valley in the USA saw many small software development and start-up companies make big in the success arena; for example Junglee, a small start-up company with a capital

of less than $40 000 thousand started by four engineering graduates from IIT in India was sold to Amazon.com for more than $10 million after a few years. Similar success stories of Indian multi-millionaires in Silicon Valley are plentiful. Many of the successful strategies owe their origin to the relentless pursuit of innovation in the software development area. One has to note that firms like Microsoft, Intel, Sun Systems and others arose as small start-up firms and proved their role in the hypercompetitive world.

Sustaining competitive advantage

Attempts to improve product quality and upgrade it constantly are the major sources of sustaining competitive advantage. For Indian business to compete in the world software market today this is one of the most important strategies to be considered. Some international examples may help in setting the appropriate strategies. Consider the consumer electronics industry in the USA and the world today. Sanyo, Sharp and other Japanese companies competed initially on costs and prices in selling portable TV sets. As they penetrated the USA and other foreign markets, they gained substantial economies of scale and reduced costs and prices further by exploiting learning-curve effects. Today the Korean competitors such as Samsung, Gold Star and others are following the same route and competing on costs and price. There is no reason why Indian companies cannot follow the Japanese and Korean style in the field of software markets.

The current value of Indian software exports is less than 3 per cent of the US software market. In 1981 it was barely 0.18 per cent, rose to 0.82 per cent in 1990–91, then to 1.70 per cent in 1994–95. By contrast, the NICs in Asia – for example Korea, Singapore, Thailand and Hong Kong (China) – have outperformed the Indian record significantly. Three characteristics of the Indian software exports sector are important to note, as they would determine the power of sustaining competitive advantage in the future. The first is the export profile. Dividing exports into three categories – software services, software packages and data entry – India's exports have mainly concentrated on the software services, as high as 90 per cent in 1992 by the World Bank's estimate. Singapore and even China concentrated on software packages to the extent of 58 per cent and 56 per cent respectively. This shows the vulnerability of India facing stiff and increasing competition in the future, in particular from Singapore and China. Secondly, the Indian software export market has been dominated by the two largest exporters – Tata

Table 3.1 Major Indian software exporters

Rank	1985/86	1989/90	1994/95
3	PCS	COSL	Wipro
4	Hinditron	Datamatics	Pentafour
5	Infosys	TI	Infosys
6	Datamatics	DEIL	Silverline
7	DCM DP	PCS	Fujitsu
8	COSL	Mahindra-BT	DEIL

Source: *World Bank Reports: Technical Report on India.*

Consultancy Services (TCS) and Tata Unisys Limited (TUL) – so that no other Indian company managed to gain more than a 5 per cent share in the 1990s. Those companies which make up the next six largest exporters after the Tatas were the strongest at the start, but failed to sustain their competitive advantage. Table 3.1 based on Heeks (1996) shows this very clearly.

The third characteristic of the software export market in India is the significant loss of skilled manpower. The impact of brain-drain to overseas competitors in the USA and Europe is quite significant and increasing over time. Oracle, Microsoft and other US companies are actively head-hunting in India for skilled software engineers and technicians. According to one estimate, about 13 per cent of Indian software developers working on exports have been leaving their Indian companies every year to go abroad permanently or semi-permanently.

It is clear that Indian software developers must forge a dynamic global strategy in order to develop and sustain their competitive advantage. The government has to play a more active role here as in Korea and Taiwan. Taiwan's example is noteworthy in that it has many electronic firms in software development which started small but shared access to the R&D results from the government-affiliated laboratories in the Industrial Technological Research Institute (ITRI). Very often the research team members of ITRI leave to set up businesses on their own, or join private companies in their R&D department. Thus, Acer of Taiwan sells under its own brand because it has its own PC sets and is hence able to offer new products and services directly competing with major US and Japanese rivals. Indian companies have yet to set up any close links with US technology institutes and research centres. Also, private firms in India scarcely contribute to R&D activity in software development in government research laboratories.

Fostering effective alliances as a global strategy

Strategic alliances between Indian software developing companies and global partners in the USA, Europe or Japan offer substantial advantages in forming a global strategy. These alliances usually take the form of long-term agreements between firms operating in the world market which attempt to exploit the benefits of network externalities, economies of scale through cooperation in component production, assembly of particular models and reducing costs of marketing. The most important benefit from alliances of Indian software developers with their US counterparts is to spread risks in developing new software and make them suitable for international markets. A classic example occurs in the pharmaceutical industry, where firms have successfully entered into cross-licensing agreements on new drug discoveries in order to hedge the risks that their own research may prove to be unsuccessful. Indian software companies have two unique roles in this global strategic alliance. Trial runs, simulations and product profiling of software packages developed in the USA may be attempted in India at a substantial cost advantage due to the skilled manpower advantage in India. Secondly, Indian developers may use the global alliance to develop their own R&D sector in both hardware and software production. Most of India's computer exports today have been basic PC units, which are assembled in India and have a very high import content. However, such exports make very little positive contribution to India's balance of trade or its overall GNP growth. On the negative side such exports tend to hinder the efficiency of technological capability in developing both software and hardware production in India. Another point is that the majority of India's hardware exports went to Russia and other former East European bloc countries under bilateral trade agreements that were paid for in Indian currency. But Indian companies had to pay for the components in foreign exchange, thus these exports caused India a substantial foreign exchange loss.

Clearly India has to modernize its export strategy in two major directions. One is to develop an effective system of alliances with US counterparts so that appropriate market niches can be set up where Indian skills have a comparative advantage. This calls for moving away from selling manpower services to developing software packages, so that trial experiments can be performed on the latter in India and unit costs can be lowered. Secondly, the alliance must not be viewed as a short-term measure to boost exports through lower labour costs, nor should they be looked on as a subsidiary of a foreign partner or as their managing agent, since these arrangements hinder the efficiency of the local industrial structure suitable for competition in the world's software markets. Too often Indian companies have opted for the easier route of agency or overseas subsidiary, for example HCL-HP subsidiary, and shifted away from developing the R&D sector and the component parts of the hardware production. Clearly the government has a very important role here in helping software companies in India, as the very successful examples of Korea, Singapore and China show.

Comparative advantage

The principle of comparative advantage has three facets of application. In the economic theory of international trade, a country has a comparative advantage in the good that is relatively intensive in the country's relatively abundant factor. Hence if labour is relatively abundant in a country relative to capital, then it should export labour-intensive goods to other countries which are either capital-intensive or less labour intensive. Similarly, a country has a comparative disadvantage in the good that is relatively intensive in the country's relatively scarce factor. This means that a country should not export goods which are intensive in the scarce factor, for example capital. Finally, the theory of comparative advantage has a dynamic counterpart which may sometimes run counter to the static aspect of the theory. As we mentioned before, software exports from India take three forms: (a) the export of software services through consultancy, overseas assignments or temporary brain-drains, (b) the support of software packages through developing and testing software of different forms in India for US and other overseas companies, and (c) electronic bookkeeping, data entry and storage services. All these forms of software exports are highly labour-intensive information services and India faces stiff competition from the following eight countries (Table 3.2) as identified by World Bank reports.

It is clear that India's competitors rely more on software packages than services. Since software package exports provide more stable, long-term and greater earnings than software services, India's skewed pattern of software exports exposes its vulnerability in the future due to its non-diversification. The relative efficiency in software diversification in exports may be measured by the percentage decline in software services and the increase in software package development. Measured by this

Table 3.2 Pattern of software exports from India and its competitors, 1990

	Proportion of exports (per cent)			
	Software services	**Software packages**	**Data entry and storage**	**Total**
India	90 (88)	5 (7)	5 (5)	100 (100)
China	17 (12)	56 (60)	27 (28)	100 (100)
Singapore	25 (20)	58 (62)	17 (18)	100 (100)
Ireland	65	21	14	100
Mexico	53 (50)	32 (35)	15 (15)	100 (100)
Hungary	40	59	1	100
Philippines	39	20	41	100
Israel	19 (15)	76 (81)	5 (4)	100 (100)

Note: The numbers in parentheses denote the percentage for 1997.
Source: World Bank *Annual Reports*.

efficiency, India's performance has been below the average. Over a decade India's dependence on software package exports has been less than 3 per cent, where even China and Mexico performed much better. However, compared to China and Mexico, India has several points of comparative advantage – more educated engineers, proficiency in the English language and links with computer and software firms in Silicon Valley through entrepreneurs of Indian origin.

Due to its dependence on software services in exports, India will face increasing competition in the international software market unless new strategies are forged by India. The relative success of Taiwan may provide some examples along this line. First of all, Taiwan has attempted to exploit the licensed clone market in different niches of the computer software market by entering into alliances and joint ventures with other international firms; for example Umax Data Systems in Taiwan working for Apple Mac OS, and Mosel-Vitelic in joint venture with Siemens AG of Germany. Secondly, Taiwanese computer and software industry are highly decentralized into networks of small firms, each being free to work for any client. The Taiwanese software and PC industry has a success rate comparable to the successful NICs in Southeast Asia. With respect to foreign investment Taiwan offers some mild selective inducement, offering subsidies up to 2.5 per cent. It also fosters active participation of small electronic and software firms into the R&D network of government-affiliated laboratories in the Industrial Technological Research Institute (ITRI). This two-way interaction is most helpful for the continued success of Taiwan in its export diversification strategy, and India has much to learn from this experience. The link between the research laboratories and private-sector firms in the computer and software development fields is almost negligible in India, and a similar state of affairs prevails in the field of technical and management education in India. Support from private-sector companies for technical and managerial research in ITTs and IIMs in India has been almost insignificant, and it is no wonder that graduates from these institutions, to the extent of 70 per cent or more, migrate to the USA and other overseas countries including Korea and Singapore. Clearly this highlights the need for a modern outlook for a new education policy at the national and state levels.

Core competence

Core competence rather than market power has been identified by Prahalad and Hamel (1994) as the basic cornerstone of success in the hypercompetitive world of technology-intensive industries today. Core competence has been defined as the collective learning of the organization, especially learning how to coordinate diverse production skills and integrate multiple streams of technologies. Four basic elements of core competence are: learn, coordinate, integrate and innovate. One has to assess in this framework the Indian software industry and the network of hardware, software and web development. Dynamic strategies need to be developed in each of these areas in order to achieve success and sustain it in today's technology-based world.

Learning has three components: intrinsic, extrinsic and teamwork. Intrinsic learning is internal to the organization, it helps intra-firm growth of knowledge and competence. Company R&D, on-the-job training, and incentive systems to explore technical know-how available in the industry as a whole comprise intrinsic learning. On a rating schedule of low, medium and high, Indian software companies fare very low in this rating scale. Extrinsic learning involves learning from the experiences of other firms and the trends of technology in the computer field. Alliances, collaborations and joint ventures provide direct access to this form of extrinsic learning, and India's rating is somewhere below the medium. It is true that many US companies have set up some collaborative arrangements with Indian companies – for example Texas Instruments, Citibank, Siemens and Unisys have helped create many of the smaller Indian software companies to facilitate exports – but most of the collaborations involve an Indian company acting as a local agent for a foreign supplier's software packages and undertaking sales, marketing and training support. This is much like the old managing agency system where firms acted as local agents for British industries in the colonial era. By 1996 most of the largest US hardware manufacturers had some form of agency link-up with an Indian software company, but apart from Novell and Oracle no major US or European software product company had made a direct production investment in India. It is only very recently during the last three years that top-notch global IT majors like Microsoft, Motorola, Oracle, Cisco, Texas Instruments, Alcatel and others have invested in Indian laboratories to develop their next-generation products. Similarly, out of a dozen top software labs across the world rated at level 5 (i.e., the highest) by SEI-CMM standards, five are in India with nearly ten at level 4. More than 200 Fortune 1000 companies outsource their software needs from India for their core operations, but few of these are joint ventures. In most cases the foreign companies use the Indian firm as a source of software labour to service its own needs or those of its customers. Hence the learning process for Indian companies does not involve any teamwork, nor does it help Indian software companies develop their own packages and market them abroad.

Coordination of diverse production skills and various innovative expertise related to the knowledge capital is central to software development and affects every branch of the modern industrial age; every new software has to be adapted to the needs of specific industries which may range from the Internet, three dimensional graphics, telecommunications, banking and manufacturing, to video games, special effects imaging, and so on. Coordination essentially involves finding appropriate niches for Indian companies where they may have dynamic comparative advantage and gather the requisite talents from India to develop new software packages suitable for international markets. Consider how US companies pressure their government to satisfy their coordination goals. The 21st century Technology Resources and Commercial Leadership Act which Senator John McCain brought to the US Senate in late 1999 was designed to keep the US high-tech industry on top by filling the need for skilled technology manpower. There is a clear need for an Indian government policy that attempts to modernize the Indian economy to recreate Silicon Valley's business culture in India. Recently, for the period 1998–2001 the Securities and Exchange Board of India (SEBI), India's equivalent of the US Securities and Exchange Commission, recommended a number of progressive steps for the Government of India: (a) establish government-sponsored incubators; (b) allow a wider array of financial institutions to attract venture capital (VC) from abroad; and (c) liberalize the complex regulations and tax laws that hinder the growth of the VC industry. Currently, Indian tax laws are friendlier towards foreign venture capital than the Indian counterpart, hence the proliferation of VC firms that are registered in Mauritius with whom India has a tax treaty. However, these steps are yet to be implemented, in practice.

One of India's great attractions is the talent pool of engineers and skilled manpower. Skilled entrepreneurs can leverage this relatively inexpensive talent pool to build businesses whose market is the whole world. Here is the need for coordination and upgrading, the other two key elements of core competence. To find the appropriate niche in the world software market and then integrate the available talent pool so as to upgrade the software packages for the world market should be the goal of Indian software companies in world markets today. Two recent examples offer hopeful signs for the future. One is Infosys Technologies which is the first Indian IPO (initial public offering) in the US NYSE market. Infosys develops software, turnkey projects and software services for the banking, telecommunications and manufacturing industries. Its shares were priced initially at $34 each but soared to $179.50 on 11 October 1999, and the successful record continues. Another example is Satyam Infoway, India's second largest internet service provider. Satyam launched in November 1998 after the deregulation of the ISP market in India. The company offers dial-up internet access, e-mail and webpage hosting to consumers in India through online registration and user-friendly software. Most importantly, Satyam has formed strategic partnerships with US companies such as America Online's Compuserve Network Services, Sterling Commerce and Open Market. Through these partnerships Infosys can offer all relevant business information and content sites tailored to Indian interest worldwide.

These examples need to be multiplied in view of India's talent pool in the PC and software sector today and the high rate of success of Indian entrepreneurs and venture capitalists in Silicon Valley in the USA. The example of Japan may provide a pointer; Japan not only borrowed US technology but improved it significantly and then competed very successfully in the US market. This is as much true in the automobile market as in the computer and related industries in the world today. In may ways India has a more superior resource base, but it needs to develop core competence, and the time is now. All that is needed is the will and determination. The will of the people and Indian entrepreneurs, and the determination of government policy-makers. As Peter Drucker (1999) points out, in developed countries today the central challenge is no longer to make manual work productive; the central challenge will be to make knowledge workers productive. Knowledge workers are rapidly becoming the largest single group in the workforce of every developed country, already comprising 40 per cent of the US workforce. It is on their productivity above all that the future prosperity and indeed the future survival of the developed economies will increasingly depend. India has a comparative advantage in the talent pool of knowledge workers, in that they provide the most valuable asset to build core competence. This is the greatest managerial challenge before India and its policy-makers.

Growth synergy in India

In order to assess the export contribution of the software industry in India one has to adopt a broader perspective. Four types of criticism have been made about the development of the PC industry and the software sector in India. One is that it has not helped the masses at all, neither has it contributed much to raise the average level of living since it is mostly urban centred and geared to the world export market. Secondly, the impact of software industry growth on other businesses and other sectors of the economy has been nil or almost negligible. It has not helped investment to grow in other sectors. Thirdly, it has accentuated the so-called 'brain-drain' movement resulting in a large-scale exodus of skilled talent from engineering and management institutes. This has resulted in significant costs to the national economy, since the engineering and management graduates migrating abroad receive large subsidies in their education and training. Finally, very few expatriates who make it big in Silicon Valley return to India or invest heavily to improve export earnings for India.

To assess the effect of the growth of software and hardware industry on the average level of living one has to look at the history of the computer industry in India. Though India was initiated into using mainframes and data machines as early as the 1960s, the Indian software industry only took off in the mid-1980s. Its phenomenal growth can be seen from the fact that the total industry output increased from Rs500 million in 1988–89 to Rs159000 million in 1998–99. This is more than 300 times the initial output level. Heeks (1996) estimated the Indian software export growth rates as shown in Table 3.3. As the table shows, export growth rates for the Indian software industry are impressive, though Indian software exports formed less than 0.15 per cent of the total world computer services and the software market in 1994–95. This

Table 3.3 The growth of Indian sofware support

	Growth in exports (per cent)	Growth in US market share (per cent)
1988–89	29	−5
1990–91	24	9
1991–92	33	18
1992–93	26	10
1993–94	43	25
1994–95	53	28
1998–99	60	35

Source: Mani (1998).

shows two things. One is that the Indian software industry has great scope in increasing its share of the world market, and the domestic multiplier and accelerator effects of such increase in export volumes would be considerable. Secondly the volume of domestic employment creation consequent on such export-led growth is likely to be very significant. The latest software developments are attracting increasingly large interest and investment in IT-enabled teleworking services like call conferencing, medical and legal transcriptions, managing large databases, web-content creation, animation and visual imaging. Companies like Swissair, GE, Bechtel, US insurance companies, the US fashion industry and Hollywood studios have outsourcing arrangements with Indian companies. Currently these subcontracting arrangements employ more than 25 000 people, and over the next eight years it is expected to create at least a million jobs. Indian software companies have demonstrated their economic and technical strength in the main domains of (a) banking and financial services, (b) insurance, (c) manufacturing and (d) internet and e-commerce. Two other areas that may be successfully explored are (e) video games and electronic toys, and (f) herbal medicines and related pharmaceutical products.

The impact of the software sector on the rest of the economy is also changing very rapidly. For the years 1998 and 1999, for example, Indian software companies sold $3.9 billion worth of software, of which almost 70 per cent was exported according to estimates of India's NASSCOM (National Association of Software and Services Companies). Their estimates further show that during the six-month period April–September 1999, Indian software exports amounted to $1.87 billion or about 10.5 per cent of the country's total exports. The major problem before the industry is how to change from the lowest component in value chain, such as offering on-site professional services or body-shopping at client's locations overseas to higher components in the value chain such as packaged products and superior consultancy services and turnkey projects.

The brain-drain argument has several flaws on the economic side. Certainly it is true that engineering and management graduates in India are trained at the Indian taxpayers' expense, which is mostly unrecovered if they seek employment abroad. However, if the domestic industry provides no scope of employment, these graduates would contribute at a very low level. Two possible remedies are to develop privately-sponsored engineering and management institutes, and foster the growth of software firms so that the tempo of employment creation could move at a faster rate. This calls for more investment in the software industry sector and active government support to encourage such investment. One area where more investment is urgently called for is the R&D field. According to estimates by *The Economist* (1994), US software companies spend an average of 14 per cent of sales on R&D. In India the figure is 3 per cent and most of it is due to government support. A recent estimate by Mani (1998) exhibits following pattern shown in Table 3.4.

One effective way to utilize the talent pool in India is to speed up research spending investment in both public and private sectors. The managerial challenge is much more on the private sector, since it can gain more directly from such investments which will raise the export drive to higher products in the value chain. On its part the national government in India can help found a 'Brain Trust', where the Indian expatriates of Silicon Valley may contribute along with the private sector. This trust could set up new training and research institutes in India, training students how to embark on new ventures in software markets in India and the world. This would counteract the steps adopted by Microsoft and Intel to drain India of its brains and talents for decades.

Table 3.4 Average real rate of growth of in-house R&D expenditure (per cent per year)

		Public sector	Private sector
1	Average of the pre-reform period, 1987–91	6.07	7.99
2	Average during the post-reform period, 1992–95	5.05	10.28
3	Crisis year, 1991–92	1.81	0.91

Source: Lale (1999).

One has to mention in this connection the recent drive by the CEO Azim Premji of the info-tech giant Wipro to bridge the huge digital divide between the technology-capable and the illiterate poor in India. It is time for other software industries to participate in this innovative venture which could offer great hope of creating the social and economic infrastructure needed in the twenty-first Century.

The criticism that few expatriates and non-resident Indians in the USA return to India or set up investments in the software industry is not appropriate or valid. For one thing, recent times have seen collaborative arrangements between these two partners, one is in Silicon Valley and the other in Mumbai and Bangalore. The main culprit seems to be the lack of a clear focus in the liberalization and industrial policy reform by the Indian government. The openness in trade required for export-led growth is much less compared to Singapore, Korea, Taiwan and even China. The following estimates (Table 3.5) by Lall (1999) reveal the overall pattern of export growth. Note that recent Indian growth is much lower than for China or Latin America.

The contrast with China – most similar to India in terms of market size and industrial structure, a historic inward-orientation and aversion to foreign investment – is particularly noteworthy. China went through several stages of 'combining plan with the market' before adopting the current goal of a 'socialist market economy'. No-one has attempted to quantify the contribution of the expansion of foreign trade and foreign capital inflows to China's economic growth, although it is known to be substantial; for example its GDP growth was 9.9 and 10.5 per cent during 1985–89 and 1990–94, respectively, and it has sustained its export growth along with GDP growth maintaining a huge balance of trade surplus with the USA. Since 1978 China has increasingly been able to import the most up-to-date capital equipment produced in Japan, the USA and Europe; the increase in the quality of capital more than offsets the slight reduction in the rate of increase in the quantity of capital.

Table 3.5 Annual growth rate of manufactured exports, current US$ (per cent)

	1980–85	1985–90	1990–95	1997
India	4.9	18.2	12.8	3.0
Mature NICs	7.3	17.9	12.4	2.4
China	–	23.1	22.4	21.0
Latin America	–	4.5	20.6	12.5

The pressure to export more and more manufactures and also software products led Chinese enterprises in these sectors to rapidly upgrade the quality of products and reduce costs. These improvements spread quickly to the production of items for both domestic and foreign consumption. Thus, according to World Bank estimates, the growth of total factor productivity (TFP) which measures technological progress attained a value of 2.9 per cent per year on average during 1990–94; it was 2.2 per cent during 1985–94. This is a remarkable rate of progress far exceeding that of the USA and NICs in Asia. One has to note that the major contribution of market reforms in China was not to the large state-owned sector, it was the small and medium-scale enterprises owned collectively or privately that benefited the most. This remarkable success record is also shared by Taiwan, where small and medium-sized enterprises spearheaded the growth and high levels of education and skill fueled the rapid growth process. For Taiwan the decentralized reforms helped reduce corruption at government level and generated an egalitarian income distribution.

The rapid growth episode of China owes a large part to the growth of its export trade, where the software industry has contributed to a significant degree. Today the Chinese economy looks very different. As Zerega (2000) points out: 'Once almost as common as bicycles, Communist Party billboards in Shanghai are being replaced by catchy dot-com advertisements urging the nation of more than 1.2 billion inhabitants to get online.' 'Sister, get on the Web!' screams a billboard sponsored by a portal and e-commerce company named pAsia. 'Unity is power' declares another. 'Brother, cut the price' is another. The slogans are paired with images of handsome youths striking militant poses and the message is sledge-hammer simple (*Red Herring*, special Asia issue no. 83, pp. 120–32: 'What would Mao think?').

China has developed the conviction and the determination that its future depends on its successful entry into the technology-intensive global economy. India needs this determination. It is the greatest challenge before Indian management which has to realize that success in today's world comes from good and innovative management and careful investment of capital, and not the sudden flash of a brilliant idea.

Concluding remarks

The greatest challenge before India today is how efficiently to enter into the software markets of the world. New paradigms of hyercompetition demand that Indian businesses follow the three Cs of economic efficiency, culminating in core competence. India has all the requisite resources – talent pool; appropriate contact with modern technology; facility in English language and communication; and a democratic form of government. It needs will and determination; the rest will follow.

References

Dahlman, C. and Nelson, R. (1995) 'Social Absorption Capability, National Innovation Systems and Economic Development', in B. Koo and D.H. Perkins (eds), *Social Capability and Long Term Economic Growth*. New York: St Martin's Press.

D'Aveni, R.A. (1994) *Hypercompetition: Managing the Dynamics of Strategic Maneuvering*. New York: Free Press.

Drucker, P. (1999) *Management Challenges for the 21st Century*. New York: HarperCollins.

Heeks, R. (1996) *India's Software Industry*. New Delhi: Sage Publications.

Jorgenson, D.W. and Stiroh, K.J. (2000) 'Raising the Speed Limit: US Economic Growth in the Information Age', in W.C. Brainard and G.L. Perry (eds), *Brookings Papers on Economic Activity*. Washington, DC: Brookings Institution.

Knudsen, C. (1996) 'The Competence Perspective: a Historical Review', in N.J. Foss and C. Knudsen (eds), *Towards a Competence Theory of the Firm*. London: Routledge.

Kraemer, K. and Dedrick, J. (1994) 'Payoffs from Investment in Information Technology', *World Development*, vol. 22, pp. 1921–31.

Lall, S. (1999) 'India's Manufactured Exports: Comparative Structure and Prospects', *World Development*, vol. 27(10), pp. 1769–86.

Mani, S. (1998) 'A Curmudgeon's Guide to Economic Reforms in India's Manufacturing Sector', *Economic and Political Weekly*, vol. 33(50), pp. 3257–72.

Norsworthy, J.R. and Jang, S.L. (1992) *Empirical Measurement and Analysis of Productivity and Technological Change*. Amsterdam: North Holland.

Porter, M.E. (1990) *The Competitive Advantage of Nations*. New York: Free Press.

Prahalad, C.K. and Hamel, G. (1990) 'The Core Competence of the Corporation', *Harvard Business Review*, vol. 66, pp. 79–91.

Prahalad, C.K. and Hamel, G. (1994) *Competing for the Future*. Cambridge: Harvard Business School Press.

Sengupta, J.K. (1998) *New Growth Theory: An Applied Perspective*. Cheltenham: Edward Elgar.

Sengupta, J.K. (2001a) 'Efficiency under Hypercompetition in the Computer Industry.

Sengupta, J.K. (2001b) 'A Model of Hypercompetition', forthcoming in *International Journal of Systems Science*.

Sengupta, J.K. (2001c) 'Level and Growth Efficiency under Technological Change'.

Shapiro, C. and Varian, H.R. (1999) *Information Rules: A Strategic Guide to the Network Economy*. Boston: Harvard Business School Press.

4
Powered by India

Mohanbir Sawhney and Deval Parikh

The Internet is the fifth network of the modern age; the first four networks were the telegraph, the railroad, the telephone, and the electric network. Each of these networks had an enormous impact on how people live, work and communicate, and the Internet is no exception. The creation of a ubiquitous network and digitization of information is leading to unprecedented innovation and opportunity, although the waves of innovation have only recently started to hit Indian shores. Over $5 billion of venture capital has been poured into Indian dot.coms and information technology companies in the last few years. With the rise of ISPs like Satyam Infoway, portals like rediff.com, Indya.com and 123India.com, and e-commerce firms like indiabulls.com, fabmart.com and clickforsteel.com, and consumer-to-consumer commerce sites like baazee.com and bidorbuy.com, the Internet is clearly coming of age in India. Indian entrepreneurs have a once in a lifetime opportunity to participate in the enormous wealth creation that the Internet has unleashed. However, they must focus on the right opportunities.

Barking up the wrong tree?

While the Internet in India is big, it may not be big enough to support e-commerce sites focused on the domestic market. International Data Corporation (IDC) estimates that there will be 2 million Internet users by December 2000 in India, growing to about 20 million users by 2004. However, the size of the Indian e-commerce market pales in comparison to global e-commerce markets. IDC estimates suggest that by 2003, developed countries will account for 89 per cent of global business-to-consumer (B2C) e-commerce revenues, and 97 per cent of the global business-to-business (B2B) revenues. The domestic market will support a few portals, a few B2C and B2B e-commerce sites, but a large majority of Indian e-commerce sites will not make it. At a time when even e-commerce leaders in the USA like Amazon, Webvan, eToys, Freemarkets, Ventro, and Internet Capital Group are struggling to reach profitability and generate liquidity in the well-developed and large US market, focusing on the Indian e-commerce market may be a risky proposition. Not only is the Indian e-commerce market much smaller than the USA or Western Europe in terms of buying power and customer base, India also needs to overcome infrastructure barriers like limited PC, Internet and telephony penetration in households and businesses.

We believe that Indian e-commerce and technology firms should focus on export-led growth, instead of growth-led exports. Traditionally, Indian industrial firms have focused on the domestic market, with exports being an afterthought, and only recently have export-oriented firms begun to achieve prominence, led by the garment and software industries. In contrast, countries like Japan and Singapore have always focused on the outside world as their primary market. In e-commerce, Indian firms should avoid falling into the domestic market trap. Like the proverbial frog in the well, Indian e-commerce firms may miss the big opportunities if they don't look outwards to the world market.

Getting virtual: the e-services frontier

So where are the opportunities, if not in e-commerce and not in the domestic market? We believe that the biggest opportunities for India-based companies may lie in the area of services that are *powered by human capital*, and *delivered over the network* to remote locations. India can become the world's leading supplier of human intelligence accessed over the Internet on an as-needed basis – a human 'utility company'. The reduced transaction costs as a result of the Internet allow businesses to outsource many of their non-core business processes like billing, order processing and customer services; and even some of their primary processes like engineering, design and development. Theoretically, a firm could have its R&D done in Silicon Valley, its engineering done in India, its manufacturing in China, sell the product in the USA, and support the customers through call centres in India or Ireland.

In a connected world, intelligence can float freely like molecules in the ether, coalescing around temporary problems whenever and wherever needed to solve customer problems. In the knowledge economy, firms need knowledge and intelligence to compete, but where the intelligence is physically located may be irrelevant. Consider computing and software as an example. In the absence of a network, software and computing power need to be located within a company's boundaries. But in the networked world, computing and software have become services that can be delivered over the Internet. This trend is evident in the rise of applications service providers (ASPs) like Exodus and Corio; infrastructure service providers like Loud Cloud and Jamcracker; as well as Dell's computing utility DellHost which offers its enterprise systems and storage products over the Internet. In the case of Dell, the functionality that was once packed into Dell's computers can now be delivered over the network, much like utility companies deliver water, natural gas and electricity over pipes. When utility companies did not exist, or were unreliable, corporations used to build and operate their own power plants and water wells. And consumers owned propane gas cylinders, as they still do in India. But when utility companies become reliable, corporations no longer need to own power generators and consumers no longer need to own portable gas cylinders. Similarly, as the Internet becomes reliable, Dell's customers may simply buy computing, not computers.

We define these outsourced activities that are made possible by the Internet as e-services. These network-delivered services can span a wide range of sectors, and may require differing levels of human capital. For instance, Exodus does not use as much human capital for every dollar of revenues it generates, when compared with an e-learning firm that provides tutoring services over the Internet. The most compelling opportunities for Indian entrepreneurs lie in e-services that involve a significant level of human capital, and can be delivered entirely over the Internet. Examples include teaching, design, architecture, software development, engineering of all kinds, statistical analysis, advertising, legal services and marketing services, to name a few. These services are quite different from the traditional notion of 'body shopping', which involves sending IT professionals overseas to staff teams at client sites. The opportunity in question is 'brain shopping', where the brainpower is delivered over the network, while the person is still physically located in India. The network allows the brain to be decoupled from the body, and it allows human intelligence and knowledge to be brought to bear on geographically remote problems.

Recently, Michael Dell asked one of the authors if he knew of a firm in India that could do customer service for Dell's small and medium-sized business customers. With the cost of telecommunications coming down, and the labour and intelligence arbitrage opportunity, business leaders in the USA are thinking of sourcing services from India. General Electric already has a 3000-person customer-care operations centre in India. In order to satisfy customer needs, firms all over the world need to source from the best possible offerings at the best cost. Hence, business leaders in India need to look outwards towards the world, and not inwards in order to profit from the opportunities that the Internet presents.

Understanding the e-services opportunity landscape

The opportunity landscape of e-services can be understood in terms of a simple set of two dimensions (see Figure 4.1 below); any service business that requires some amount of intellectual capital can be put on this matrix. The first dimension of the matrix is the *human capital intensity* of the activity, which relates to the quantity as well as to the quality of the human capital. For instance, medical transcription is a manual process, but it requires a relatively low human skill level. In contrast, management consulting services provided by the likes of McKinsey or BCG require a high level of human capital. The second dimension is the *degree of digitizability* of the service. Some services can be delivered satisfactorily over the network; for instance, services that

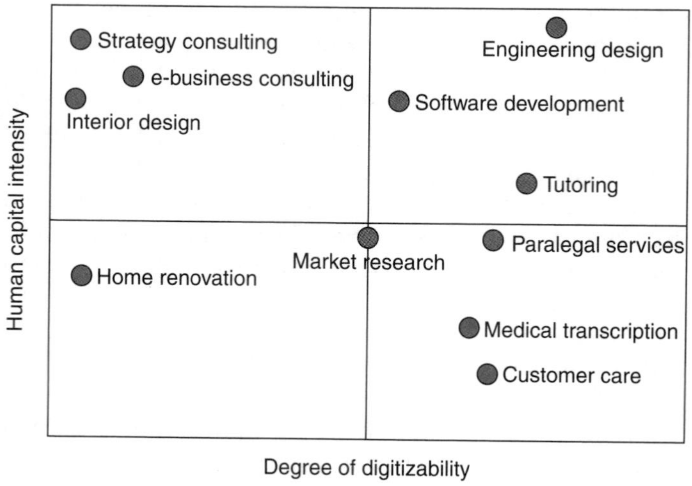

Figure 4.1 Structuring the e-services landscape

involve digitized information like software code, AutoCad files and data are intrinsically digital in nature. Similarly, customer support services are quite amenable to digitization, and can be delivered over the network. However, management consulting and interior design are 'high-touch' services that require a high degree of personal interaction with the client.

The greatest opportunity lies in the upper-right quadrant, where the human capital intensity is high, and so is the digitizability of the service. The first dimension assures the *relevance* of the competitive advantage of Indian human capital; the second assures the *portability* of this human capital over the network. Essentially, whenever brainpower is needed in plenty for an e-service, and whenever this brainpower can be accessed remotely over the Internet, there is an opportunity for India-based e-service providers.

Consider the field of engineering of mechanical and plastic parts that uses techniques like computer aided design (CAD) and finite element analysis (FEA). Using the latest software tools and specifications from customers, design shops in India can design the latest and greatest components and systems for the world's leading companies. The field of education also holds a lot of promise. In the USA hundreds of thousands of students in high school and in the first two years of college go through classes in mathematics and the sciences. In the average public school or college, students need a lot of hand-holding to understand these new domains of science. Educational institutions can only provide limited tutoring services, because of the prohibitive cost of tutors' time. India, on the other hand, has a vast resource of well-qualified English-speaking teachers. The Internet allows students in the USA to source knowledge via interactions with Indian teachers at a fraction of the cost of a US-based teaching assistant.

India can play a winning role in the emerging arena of e-services by leveraging its vast reservoir of grey gold – its intellectual horsepower and its 'installed base' of millions of English-speaking professionals. India-based entrepreneurs can create interesting Internet-based service businesses that function as human services providers (HSPs) – the human equivalent of application service providers. These opportunities span a wide range of areas, including product development and design, contract R&D, market data analysis, software modelling, e-learning, and customer services for customers in developed countries. There may also be attractive opportunities for firms to manage several business processes for a single client in an integrated fashion, by aggregating several HSPs and taking responsibility for their performance.

Getting personal: custom-manufactured products

Firms all over the world are going to great lengths to provide customers with personalized experiences and customized products. Even cosmetics and t-shirts can be customized over the Internet and delivered to the home. In many cases, the actual manufacturing of the customized product is automated, but in other cases true personalization requires personal attention and personal effort in the manufacturing process. This is particularly true when the product in question is hand-crafted and involves significant human effort. The arena of custom-manufactured products is another unique opportunity for India to leverage its skilled and semi-skilled worker base, as well as its intellectual horse-power. Just about any product or service where manual customization is the only way to deliver truly personalized products can fall within this arena.

Consider the garment industry. Indian artisans have accumulated centuries of knowledge on weaving the best handloom fabrics, using a wide range of designs, fabrics and styles. However, this capability has been only partially exploited because of the physical separation of customers from the artisans. A consumer in the USA or Western Europe shopping for a rug may choose a particular rug and matching pillow covers; the retailer of home furnishings can have actual samples as well as a digital catalogue where consumers can choose the dimensions and type of the rug. This information can be transferred directly to the handloom operators in India, and the finished goods can be shipped overseas in a matter of days. This process can be used to offer custom-manufactured products to customers halfway across the world.

Another example of a custom-manufacturing opportunity is Ayurvedic medicine. There is an increased awareness in the West about the benefits of natural remedies relative to conventional medicine, but, ideally, Ayurvedic remedies need to be adapted to the specific body type of the individual and to their symptoms. To create personal remedies, patients may need to consult a doctor and then have their personal remedies prepared by hand. This consultation can be done between patients in the USA and Ayurvedic doctors in India over a website, using chat or videoconferencing technology. Based on the consultation, the doctor can then have his staff prescribe a custom blend of the right medication and then ship it to the patient abroad.

We can think or a diverse range of custom-manufacturing opportunities, ranging from engineered parts, to custom furniture, to custom tailoring. Wherever there is a significant labour and skill component needed in manufacturing, and wherever the product does not have to be immediately delivered to the customer, there may be an opportunity to create a web-based custom manufacturing entity that can connect Indian skilled labour to a global customer base.

Implementation challenges

Implementing the vision of e-services and custom-manufactured products for foreign markets is not an easy task. The devil is in the details. The goal of the firms based in India should be to provide their customers with a reliable, high-quality and consistent experience, but there are four main challenges before this vision can become a reality.

Establishing business-development teams

The first step for every business should be to identify the correct market and how it can best serve the needs of its potential customers. Indian firms will need relationship managers and sales managers in the developed world to effectively sell the offering via a face-to-face meeting. There are simply no shortcuts to good selling, and having a presence abroad is an important hurdle that firms in India need to overcome. Several firms based in Israel like IDT, Net2Phone, DealTime and Gilat operate with similar structures, they have business development teams in the USA and they source the technology from development centres based in Israel. Indian firms could also leverage the vast talent pool of thousands of US-trained managers that have lived and worked in India and know how business is done in India.

Managing customer relationships

The ventures that we have identified all need relationship managers who do the front-end sales and service for the customers outside India. The links from the customer to the specific knowledge hub should be flawless and supportive of their needs; the service provided to the customer should be a seamless end-to-end experience, the entire process from order entry to the delivery of goods should be coordinated across all the parties involved in creating the customer experience.

Tools and training

Firms that participate in these e-services should have domain-area expertise, and an operating knowledge of the service that is being outsourced to India. Adequate training (or retraining) in both these dimensions is a must since becoming world-class requires world-class training. Firms should have the latest software tools that aid the productivity of their staff, and ensure compatibility with their customers abroad. They should also train their employees well in the domain, local cultural and business practices, and even in the accent, writing styles and local mannerisms.

Quality assurance

If a firm outsources mission-critical processes like billing and engineering design, the quality of the services and products is an attribute of paramount importance. Quality has to be impeccable the first time, and every time. Indian firms have a reputation of doing a great job initially, but gradually slacking in quality standards over time. Having the equivalent of ISO 9000 certification for business processes and practices would go a long way to assure foreign customers of the service quality provided by e-services firms.

Institutionalization of knowledge

The knowledge gained from the experience of carrying out several projects should become a part of the institutional memory of the firm in India. This knowledge can be leveraged for future projects, and can enhance the training and retraining of the Indian firm's employees. Additionally, the deep domain area expertise from compatible firms can be combined to get bigger projects in the future.

Conclusion

According to former American President Woodrow Wilson, 'There are many voices of counsel, but few voices of vision . . . and little concert of thoughtful purpose. It is our duty to find ourselves.' The e-services and custom-manufactured products arenas offer several advantages for India, including better wages for knowledge workers, a higher productivity of the workforce, and knowledge gained by working on world-class projects for leading global firms.

We can visualize a future where India can brand this effort at a national scale as 'Brainpowered by India'. Every analogue or digital transaction that would happen on paper or on a website would have this symbol, much like the ubiquitous symbol of 'Intel Inside' on just about every PC that is bought. This effort would put a stake in the ground that India has the talent and the intellectual horsepower that goes beyond software engineering, and would establish India as a knowledge hub for the world.

5

Using Offshore Software Development to Drive Value: The Experience of Motorola in India

Soumitra Dutta

Indian software: a success story

The Indian software industry is often quoted as one of a few bright spots in India's economy where local companies have been able to successfully compete with other global players. Both local political leaders such as Prime Minister Atal Behari Vajpayee ('IT is India's tomorrow'[1]) and global software luminaries such as Bill Gates ('India is likely to be the next software superpower'[2]) have attested to this promise of the Indian software industry.

The data released from the most recent annual industry survey of the National Association of Software and Service Companies (NASSCOM; www.nasscom.org) provides some interesting insights into the rapid growth of the Indian software industry over recent years. Some highlights from the survey regarding software exports are as follows:[3]

- Indian software exports in 2000–01 grossed revenues of US$6.2 bn and registered a growth rate of 55 per cent over 1999–2000 performance. During the year, the number of quality certified software companies increased to over 250; 27 Indian companies now have the unique distinction of an SEI-CMM[4] level 5 certification.
- Software exports increased their share in India's total export revenues. In 2000–01, Indian software exports accounted for over 14 per cent of India's total exports. In 2000–01, more than 30 software companies in India exported more than US$44 million worth of software and services; 75 companies exported more than US$11 million worth of software.
- India exported software to 102 countries around the world, with almost 62 per cent of exports to North America, 24 per cent to

> Europe and 4 per cent to Japan. During the year, one of every four global giants outsourced their software requirements to India. Offshore software development exceeded 44 per cent of total software exports (in 1991–92, offshore software development contributed only 5 per cent of the software exports).

It is evident that software export is a growing business in India. The data from the recent NASSCOM survey suggests that the global economic slowdown is in fact providing an added impetus for leading multinationals to look seriously at India for outsourcing their software requirements. Not only are US companies increasing their IT investments in India, companies from Europe and Japan are also increasing their outsourcing to India.

The strategic importance of software

Software forms the backbone of several industries such as banking, airlines and publishing, and is an increasingly important value-adding component of consumer products such as television sets, cameras, cars and mobile phone sets. Software is today a dominant force in enabling companies to exploit new distribution channels, create new products and deliver differentiated value-added services to customers. In reality, there is often little difference between an organization's software strategy and its business strategy.

In addition to the ubiquitous nature of software, the amount of software code in most consumer products and systems is doubling every two to three years. This increase is being driven both by escalating demands placed on the functionality of software systems and the rapid pace of progress in the enabling hardware technology. Consequently, software developers are scrambling to cope with the pressures of developing systems which are not only a couple of orders of magnitude bigger and more complex than those developed a few years ago, but which also need to meet ever-increasing demands for higher quality and superior performance.

Organizations worldwide face several challenges in leveraging the strategic potential of software. There is, for example, a need for a better execution of software projects; stories of dramatic time and cost overruns of software projects are legendary. Industry observers note that for every six new large-scale software systems that are put into operation, two others are cancelled and the average software project overshoots its schedule by half. The US Internal Revenue Service[5] (IRS) conceded in

1997 that despite having spent $4 billion developing modern computer systems, the systems 'do not work in the real world'.

Even organizations with a distinct history of IT achievement are not immune from such crises. American Airlines built its reputation of IT excellence during the 1980s on the back of its famous SABRE airline reservation system. However, it was the same American Airlines which had to eat humble pie in major disasters, for example while attempting to build the CONFIRM reservation system in 1992 for hotel and car rental companies such as Hilton and Budget. In 1992, which was a huge loss-making year for the US airline industry, American Airlines had to take a $165 million write-off due to the failure of the CONFIRM project. American Airlines was also the butt of several jokes in the European media when its efforts to build a reservation system for the French railway system led to a major fiasco during the traditional peak summer tourist season of 1994. Software errors caused havoc with the traditionally efficient French railways and either grossly overbooked or underbooked trains leading to a national outcry.

Offshore software development

The global volume of IT services is growing at an astonishing rate. According to the International Data Corporation (www.idc.com), global spending on IT services will grow from $439 billion in 2001 to $700.3 billion by 2005. The United States will lead all other nations in this category, spending $335 billion in 2005, up from $206.9 billion in 2001. Western Europe will spend $192.4 billion in 2005, up from 127.5 billion in 2001. Meanwhile, spending in Japan will grow from $53.2 billion to $75.2 billion.[6]

Couple this remarkable growth of IT services with the mission-critical nature of software and the challenges of effectively managing software projects, and it is not surprising to note that corporations worldwide are rethinking their software strategies. Many are now looking actively at India (and other nations) as a source of software talent. Marty McCaffrey, founder and executive director of Software Outsourcing Research based in Salinas, California, comments on the motivation for considering offshore software development:

> While cost savings were the primary reason companies outsourced offshore in the early 1990s, in recent years priorities have been changing. Timely access to highly qualified technical talent, faster time to market and accelerated delivery, ability to significantly expand

> the organization's software development capacity at minimal costs, opportunity to accelerate the improvement in their own software process and quality capabilities by working with world-class offshore companies, and reduction in the risk of cost overruns and late projects are frequently cited by customers ahead of cost savings.[7]

'Offshore outsourcing is no longer an emerging trend – it's a way of life for many IT executives.'[8] However, most do not yet feel fully comfortable with the realities of offshore software development. 'The very places where foreign development is hottest are also at times home to earthquakes, typhoons, floods and shaky infrastructures that can knock out communications and political instability.'[9] Concern about entrusting mission-critical software projects to foreign partners is real and frequent. For example, Allan Hackney, senior vice-president of information technology at Bank of America Commercial Finance (BACK) in Wilton, Connecticut, chose to use an onshore workforce from an Indian company (at a much higher cost) than to send projects offshore. His justification:[10]

> Offshore projects work when requirements are well-known and success is easily defined and understood. Onshore works better where the opposite is true: if you have an iterative development cycle or when the end point is a moving target.

Farley Blackman, CEO of StrategIM, a Burlington, Vermont consultancy specializing in offshore development adds, 'People don't understand how to work overseas, and the assumed loss of control scares some companies away from offshore development.'[11] Most firms feel comfortable shipping only niche or low-priority projects to offshore partners. The literature on offshore software development abounds with wisdom about 'Do's' and 'Don'ts' – all from the perspective of containing the risk of failure. Consequently, few corporations are exploiting the true potential of offshore software development.

The situation is analogous to the challenges faced by firms establishing manufacturing sites in foreign countries. Kasra Ferdows described it thus in the *Harvard Business Review*:[12]

> Many companies are not tapping the full potential of foreign factories. They establish and manage their foreign plants to benefit only from tariff and trade concessions, cheap labour, capital subsidies, and reduced logistics costs. Therefore, they assign a limited range of work, responsibilities, and resources to those factories.

However, this is not the only way chosen by firms. Kasra Ferdows notes that there are exceptions:

> But there are companies that expect much more from their foreign factories and, as a result, get much more out of them. They use them not only to gain access to the usual incentives but also to get closer to their customers and suppliers, to attract skilled and talented employees, and to create centres of expertise for the entire company.

Motorola in India

Motorola's experience with offshore software development in India is one such exception in the software world. It provides a unique perspective into how firms can leverage Indian software excellence for enhancing their global competitive position. Bucking the usual trend, Motorola has for over a decade assigned a unique strategic responsibility to its software development centre in India. It has invested in developing the competence of the centre and given it some of the most challenging and mission-critical projects within the entire corporation.

The payoff has been significant. Started as a greenfield site in 1991, Motorola India Electronics Limited (MIEL) stunned the software world by achieving in 1993 the highest possible software process maturity rating of level 5 in the Capability Maturity Model (CMM) (Box 5.1) nearly two years ahead of schedule. By the mid-1990s, MIEL had obtained widespread recognition in the press and the software world for being one of only two organizations worldwide with software processes certified at level 5. This was particularly noteworthy given the widespread acknowledgment of the lack of software maturity in most organizations. Given the criticality of software for Motorola's business success, the achievements of MIEL had not gone unnoticed within the corporation. In 1994, MIEL received Motorola's Chief Executive Officer's Quality Award, the first software unit to be recognized in such a manner. Mohan Kumar, the Managing Director of MIEL, noted: 'We find that we have been successful at that to the point that we produce software for about $5 b[13] of Motorola's products. This organization is in the critical path of developments. We're core to the success of the business.'

Motorola has enhanced its software excellence globally and has created a series of offshore software centres across the world to leverage the successful Indian experience. It has showed the way for other firms to make the most of their software centres in India (and other offshore locations). The following sections expand in greater detail upon the experience of Motorola in India.[14]

Box 5.1 Software process maturity and the CMM

The roots of research in software maturity assessment can be traced back to the mid-1980s when the Software Engineering Institute (SEI) at Carnegie Mellon University began developing a framework for assessing the maturity of software. From a first description in 1987, the framework has evolved over the years into the widely accepted Capability Maturity Model (CMM) for software maturity assessment.

The core concept underlying the CMM is that of five maturity levels – Initial, Repeatable, Defined, Managed and Optimized – which define an ordinal scale for assessing the maturity of an organization's software processes. At the initial level, an organization's software processes are *ad hoc* and occasionally even chaotic. In contrast, at the highest level, continuous improvement procedures enabled by the appropriate use of metrics are institutionalized within the organization's software processes. The following table provides a summary of the key process areas considered within each of the levels of the CMM.

Maturity level	Key process areas
Level 5: Optimized	• Defect prevention • Technology change management • Process change management
Level 4: Managed	• Quantitative process management • Software quality management
Level 3: Defined	• Organization process focus • Organization process definition • Training programme • Integrated software management • Software product engineering • Inter-group coordination • Peer reviews
Level 2: Repeatable	• Requirements management • Software project planning • Software project tracking and oversight • Software subcontract management • Software quality assurance • Software configuration management
Level 1: Initial	• None

The birth of MIEL

Roger Fordham, a former Managing Director of MIEL, explained the genesis of the company:

> Around the late 1980s, there was a recognition by many, not all, but many in the company that software was going to be a big part of our life. Our products were turning digital and required software. We were essentially a manufacturing and semi-conductor skilled organization. We were going to have to become equally skilled and equally driven by software as we were with those technologies. We'd better learn about this thing, better start developing competence and capability or we were going to find ourselves in trouble.

At that time, the state of the art in software development was poor in terms of quality, cycle-time and customer satisfaction. Also, there was an urgent need to fill the estimated annual shortage of 5000 staff years of software skills within Motorola; so the decision was made to set up a Greenfield site outside the USA. Mohan Kumar added:

> It was a grand experiment within Motorola. Motorola had never outsourced any software development outside the USA and there were some fundamental questions in the minds of many: Can high quality software be produced outside in a developing country? Can software be done different from what was common in the USA?

When it came to putting a Greenfield together, the Motorola Software Engineering Steering Committee adopted a 'clean-sheet' approach. Certain questions were asked repeatedly: How should we attack the (software) issues? How can we do it right? If we had a clean sheet, how would we do it? The discussions led to a decision to build a process-orientated entity somewhere outside the core Motorola organization with a high degree of management commitment and adequate resources. Sarala Ravishankar, the Manager of Quality at MIEL, gave her point of view:

> The vision came from the States, partly because of Motorola's experience with manufacturing quality and six sigma. It was thought that having some similar philosophy for software would help to improve software quality; when you start a clean sheet software factory, you should be able to start with all the disciplines and elements of the

> process in place, so that you could achieve that much faster and this could also become a learning point for the rest of the organization.

An ambitious goal was set for the Greenfield: to start at CMM level 3 from day one and to reach level 5 within four years. The goal was daunting because no-one else had attempted it before. There were also some software professionals who doubted whether it was possible to achieve level 5. Motorola did not have a large business presence in India in the late 1980s. However, the start of the liberalization of the Indian economy, the availability of a large core of cheap English-speaking engineering talent within the country, the promise of the future potential of India's emerging market and the good experiences of other American corporations (such as Texas Instruments) in setting up offshore software sites in India made it a natural first choice for Motorola's greenfield site.

Office space was rented in a hotel in Bangalore and about a dozen engineers were hired in 1991. A particular effort was made to hire graduates, even some PhDs, and often from non-software backgrounds. The initial team members were chosen very carefully, both in terms of skills and attitudes towards experimenting with new approaches. The initial recruits went through an intensive three-month training programme on various aspects of software engineering and quality management. The focus on software processes was intense. Mohan Kumar elaborated:

> When we started, we did not have any customer or any project. There was no mandate to deliver anything. There was simply a dedicated focus on understanding the process.

Jayanth Kattepur, a member of the original team, described the reactions of the team members:

> I have never found such an enthusiastic crowd of people because it was really something new for all of us, and we were trying to make the best of it. We reviewed each and every document, and discussed the relevance of every single item. Whenever required, we added our own ideas. The end result was not dramatically different, but we understood the whole process. More important, a sense of ownership was created.

George Smith, the first Managing Director of MIEL played a critical role in motivating the newly-hired engineers, and Jayanth Kattepur described his role:

> George Smith had a tremendous impact on us. He taught us. I've not seen anybody else pursue a goal with the same level of dedication. He used to come to every one of us almost every day and have a chat with us on software engineering practices. I don't think that any one of us had even an inch of doubt that this would be a success.

The growth of MIEL

By the end of the 1990s, MIEL had grown to over 450 engineers working on a new site in the heart of Bangalore. Assimilating the new employees was relatively easy as there were well-defined and operational systems for continuous improvement, training, quality, measurement and product engineering. As new employees were hired, MIEL followed one hard and fast principle: all engineers had to go through a mandatory six-week Induction Training Programme (ITP) in batches of about 20–25 employees. Roger Fordham explained the importance of the ITP to MIEL's success:

> We don't let anyone work on any of our projects unless they go through our ITP. Every one of our employees goes through that, and that gets them up to speed with our process, our tools, Motorola's way, the quality initiatives we have and the way and approaches we have, and tries to put them in a frame where when they come out they have a common language of discourse and then go on and develop software. That's a key message – we never let people go in unless they're indoctrinated or trained in Motorola's software processes.

Sarala Ravishankar, the Quality Manager, gave her view on the role played by the ITP:

> I would say the ITP is more a culture building exercise rather than a technical skills communication program. Techniques are taught to people, but more importantly we stress why are we doing what we are doing. As you start getting people to think about this, you create a fundamental belief in some of our systems.
>
> Another interesting aspect of the ITP is that it creates a certain bonding amongst the group members that lasts as long as they are in the organization. We often find ITP batch members from across projects coming together for informal lunches or anniversary parties.

MIEL achieved a CMM level 5 rating in 1993.[15] However, what was surprising was that MIEL succeeded two years ahead of schedule! By

1997, MIEL was also close to six-sigma quality (that is, about three defects per million lines of program code). This was widely recognized as a remarkable feat in the world of software engineering.[16] Mohan Kumar added:

> The success of MIEL triggered a new wave of thinking within the Indian software industry[17] and influenced the mindset of other companies that set up software centers in India over the last years.

Customer relationships and domain knowledge

MIEL only developed software for internal Motorola customers. It had complete freedom in choosing customers and it was rare for Motorola's hierarchy to push a project onto MIEL. Most of these clients had their own internal software groups, traditionally coming to MIEL to save time, reduce costs and obtain higher-quality products. However, the increasingly critical importance of software to Motorola's business and the acknowledgment of MIEL's strengths within Motorola were changing the paradigm of relationship between MIEL and its customers. Customers were coming to utilize the skills of MIEL to develop state-of-the-art software products. Roger Fordham explained:

> We have gone through three generations in our relationships with customers. Stage 1 – what do you want? You say 'I want it green, 5 by 7 by 3, and it's got to have this density.' We'll do that for you. As you evolve you come back and the customer says to you 'you know a bit more about my business now. Why don't you tell me what you think we should do?' And we have a mutual goal-setting activity. And today, we've moved on to a third generation, in which the customer is saying 'everything there is to know is known by you – so don't come to me asking me what to do. You'd better be telling me what to do.'[18] And this is exactly what we're doing. We're telling the customer what features should be done next.

The changing demands of MIEL's customers were requiring employees to develop an in-depth knowledge of their customer's markets and do their own research. Mohan Kumar added:

> The key question for us is: what is our ability to increase our depth of domain knowledge. We need to be able to proactively suggest software solutions and product ideas to our customers in Motorola.

To facilitate the need for more domain knowledge (in areas such as VLSI design and networks), MIEL had started hiring more domain experts, such as PhDs in VLSI design and networks, both locally and from other Motorola businesses. The organizational structure of MIEL was also changed to facilitate the creation of appropriate domain knowledge. Srinivas Pannala, Manager for Business Development and Strategic Planning, noted:

> Customers increasingly ask us for suggestions of software solutions to improve their products. Our ability to generate these ideas has increased by the transition to the new organizational structure.

Processes and systems

A phased development approach with parallel development and test was standard in MIEL. Each phase was managed explicitly and rigorously by a series of quality audits, built-in causal analysis and feedback mechanisms. There was a deliberate and conscious focus on extensive use of metrics. Sarala Ravishankar explained:

> We use a lot of numerical data; data on projects such as cycle times, quality, productivity and defects. We have systems in place where project data is captured, analyzed and made visible on at least a monthly basis to management. Almost everything we do is measured and the measurements are tracked visibly in a continuous attempt to move them in the right direction.

Each project team had a Software Quality Manager who was responsible for auditing adherence to key processes (such as inspections), helping the team in applying different quality tools (such as root cause analysis) and ensuring the transfer of operational project metrics to the Quality Department. The Quality department was responsible for maintaining data for all MIEL projects along with other accepted industry benchmark data. These data were utilized by project teams for estimating different aspects, such as productivity, quality and number of defects. There was a constant focus on the adoption of rigorous project management techniques during every phase of each project. Documents such as a project management plan, a quality assurance plan and a configuration management plan were mandatory for each project. Weekly status reports and monthly project reviews were used to track the progress of projects.

A formal management process was in place for handling change requests. Each project team had a Change Control Board[19] that evaluated the impact of all change requests on schedules and effort and communicated this impact to customers. MIEL tried to both educate customers and substantiate its own estimates with numerical data. Most customers viewed MIEL's estimates about costs and time with a high degree of credibility. Frequently, customers took MIEL's operational data such as those about productivity, defect frequencies and quality levels to benchmark their own software groups or their external software subcontractors. The Change Control Board was also responsible for the implementation and review of accepted change requests.

A learning environment

Learning and sharing of knowledge plays a key role within MIEL. Information about the performance of project teams was public and available to all. Project reviews were frequent and open to all. They served as a forum for all to constructively review and evaluate the experiences of a project team. Common problems were identified in these meetings and suggestions for improvements were given by all present. A manager characterized it as a 'common, practical learning environment, not just a theoretical process'. However, this culture of open criticism took some time to become second nature. Jayanth Kattepur commented:

> I remember my first review meeting. There was literally a fight because the project team being reviewed could not accept the comments given by others. It required a certain degree of openness to accept the different views of your colleagues, many of them junior to you. But we have come a long way from that today.

When MIEL was small, project reviews were attended by a majority of employees, but this was becoming less feasible as the size of MIEL and the total number of reviews (about 10 per month) increased. However, Mohan Kumar and senior managers from different functional areas reviewed every single project in MIEL on a monthly basis. There were a number of mechanisms to get employees from different functional areas of the organization together. Roger Fordham explained:

> There are lots of cross-functional and cross-operational meetings which are there partially to review the status of processes but are also to share creative solutions in things. For example, we have a

> quality meeting monthly where the quality department gives their opinion of how we're doing because in our organization quality is a management tool to point out the difference between what we're doing and what we said we'd do. And that vector between them is what quality reports on as a management tool. We also have a series of task forces and committees; so whenever management and technical folk get together in an attempt to bring out commonalties of network management architecture and functions, protocol developments, all those sorts of things, there's a weaving together of people in a variety of organizational structures to share and learn new things.

An annual event was held to highlight best practices in different key process areas, and special emphasis was put on learning from failures. Sarala Ravishankar explained:

> We have had several failures. In fact I think that failure is good for the organization because that's the best and fastest way of communicating learning to the organization. If we fail in one project, we look at what are the reasons which really caused that failure. And we have a system in place which analyses those failures and spreads the learning across the organization.

Roger also emphasized that there are mechanisms to facilitate learning across MIEL and other Motorola units:

> We also have a number of symposia. Motorola has a symposium in Europe, the US and Asia on software, generally annually, and there's a sharing which goes on within the regions and across the regions, and of course we have a number of customers. Some of our customers actually say 'We're interested in you because you have this skill. I want this out of you as a result of our business relationship.' Consequently, we often send some people over to train our customers. We work out an action plan and that'll be part of the delivery list. They'll pay us for it. That's an element of our business internally, to sell those things, not at rapid profits but to recover our costs to make it worth our while and to make it valuable to them.

By 2000, a few Motorola units had reached CMM level 4 and steady progress was also being made by many other Motorola software organizations in increasing their levels of software process maturity.

Different strategies for offshore software development

Figure 5.1 illustrates a simple matrix to position different strategies adopted by firms with respect to offshore software development. The horizontal dimension refers to the degree of competence of the offshore software development centre. A high degree of competence would be reflected in an ability to undertake more complex projects, to manage product design and development, and to proactively suggest software solutions and product innovations. The vertical dimension of the matrix refers to the strategic role of the offshore software centre within the (parent) corporation's business strategy. A highly strategic role would imply that the centre is seen as a vital ingredient for the future success of the corporation, that mission-critical projects are assigned to the offshore centre and that appropriate investments are made to support its strategic role.

From figure 5.1, four possible strategies to manage offshore software development centres may be identified:

1 *Code converters*. This is the typical strategy used by most firms with respect to offshore software development. The offshore centre is seen primarily as a source of low-cost labour. Projects assigned to the centre are labour-intensive, but do not need significant technology and design skills. Examples of typical projects in this strategy would be Y2K and system migration projects – where the functional specifications are well-defined and the offshore centre is required to execute to the stated specifications. Control for the project rests with the parent corporation and there is little empowerment for the software developers in the offshore site.

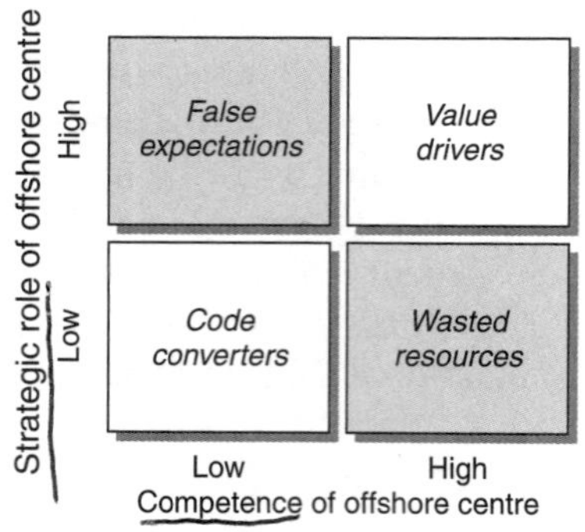

Figure 5.1 Offshore software development strategies

2 *Wasted resources*. A firm employing this strategy would make investments in upgrading the competence level of the offshore centre but not assign the appropriate strategic role to the centre within the corporate strategy. The investments may be reflected by the hiring of employees with advanced degrees, enhanced training for the offshore programmers, and increased job-rotations of these programmers to sites within the parent firm. However, the firm would be risk-averse and not assign mission-critical projects to the offshore centre. The firm would retain control of product design and innovation and rely upon the offshore centre primarily for cheap labour to execute well-defined projects. The advanced skills and training of the offshore employees would be poorly leveraged in this strategy.

3 *False expectations*. In this mode, the firm would have high expectations of the offshore software development centre and give it more responsibility for product design and innovation. However, the firm would not make commensurate investments in upgrading the competence of the offshore centre. Employees in the centre would lack the skills and training to undertake the advanced projects and the rate of successful project completion would be low. The firm may aim to emulate other successful examples (such as that of Motorola in India) but may find it impossible to do so given the low level of competence in the offshore centre. The offshore centre can develop into a drag on the overall competitiveness of the firm.

4 *Value drivers*. This is the strategy adopted by leading firms such as Motorola in India. The offshore software development centre is seen as core to the future success of the firm. Consequently, the firm makes all necessary investments in developing the right competencies within the centre – such as by hiring employees with advanced degrees and investing in quality processes and systems. The offshore centre is seen as having the ability and knowledge to innovate and create new products, processes and technologies for the company. It becomes a catalyst for change within the firm and becomes an active partner in creating value and enhancing competitiveness.

Using offshore software to drive value

It should be intuitively evident that two of the above strategies, Wasted Resources and False Expectations (shaded in Figure 5.1), are not sustainable – for different reasons. The former results in decreased efficiency for the firm; the stability of the offshore centre will eventually be questioned, as it will be unable to attract the best employees due to the low degree of challenge in the assigned projects. The latter results in increased project failures and stress for both the parent firm and the offshore centre. The parent firm will over time have to either reduce the strategic role of the offshore centre or invest further in developing its competence.

A firm could continue to use an offshore centre on a sustainable basis as a code-converter outpost. Investments in offshore employees could be minimized and the firm could assign only low-risk and low-challenge projects to the offshore centre. Most firms tend to be risk-averse with respect to offshore software development and thus tend to adopt this strategy *vis-à-vis* their offshore centres. However, firms have to be aware of one critical success factor in the code-converter strategy: it is vital for corporations to have the necessary competence in software in their home bases (given the ubiquitous presence and strategic importance of software in almost every firm's portfolio of products and services). It is both about numbers of employees and levels of skill – the challenges on both dimensions should not be underestimated. There is an acute shortage of IT professionals in all developed countries and this is not expected to become any better in the near future. A good indicator of the continuing shortage is the fact that the number of undergraduates majoring in computer science in US universities has been falling each year (from its peak of 41 889 graduates in 1986 to 24 404 graduates in 1995[20]). 'IDC projects that in the United States, the gap between the number of IT workers required and the actual projected headcount will continue to widen substantially. In the USA, this gap is expected to increase to 328 660 in 2002 from about 72 003 workers in 1995'.[21] On the skills dimension, if the CMM model rating is any indicator, firms continue to languish at the lower ends of the scale and even today most of the highly rated companies on the CMM model tend to emanate from India. If the firm lacks the appropriate levels of software competence in its home operations, the code-converter strategy will be doomed to fail.

While the code-converter strategy may be the appropriate strategy for some firms, the value-drivers strategy is the recommended strategy for most companies. There are important factors that make the adoption of the value-driver strategy an imperative for firms today. First, there is the all-important question of numbers. In contrast to the declining numbers of computer science graduates in the USA (and Europe), the number of technical graduates in India (each year) is large and increasing. According to industry estimates there are over 75 000 technical graduates from Indian universities each year. These numbers imply that the role played by Indian software professionals in the global IT industry is only going to increase in the future. At a 39 per cent CAGR, (Compound Annual Grouth Rate) the Indian IT services sector is expected to produce about US$30 billion in revenues in 2004 – this comprises about 5 per cent of IDC's global IT services spending forecast (up from 1.6 per cent in 1999) and 17 per cent of total projected IT outsourcing (up from 4.9 per cent in 1999).[22]

Further, when one looks at the evolution of IT in firms in the near future, the Internet is the decisive factor forcing firms to rethink their entire portfolio of IT services and needs. Not only is there a need for companies to change their core systems and programming, there is also the need to fundamentally rethink the use of technology for product development, design and delivery (witness the emergence of intelligent home appliances such as refrigerators and online shopping). It is no longer just a question of technology, but a question of how technology can be leveraged for innovation and competitiveness. There is a lack of professionals and employees who can help with this transition (especially within small and medium-sized firms).

Offshore software centres can play an important role in helping this business rethink. This is supported by the experience of firms such as Motorola who have invested in their offshore centres with the explicit goal of using them to stimulate product design and innovation. Indian software firms have historically not focused upon business innovation and consulting services. However, this is changing fast as they look for greater revenue-generation propositions. Indian software professionals have proven to be very adept at adapting to the changing needs of their clients and Indian software firms continue to make increased investments in R&D. According to Nasscom, R&D spending by Indian software houses touched 3.5 per cent of their total revenues in 1999–2000, and this is expected to increase steadily over the next years. Thus Indian software professionals are going to be more prepared to support the value-drivers strategy in the future.

It is important to recognize that the value-drivers strategy is probably the hardest to execute. As can be seen from the detailed Motorola case study presented above, adopting this strategy in offshore software development is not easy – a unique combination of ambitious vision, determined leadership, talented people and rigorous processes is needed. It is thus not surprising to note that other firms have found it extremely hard to replicate the success story of MIEL.

The experience of Motorola in India provides a rich set of lessons for all firms contemplating or engaging in offshore software development. If done properly the efforts can be very rewarding. Motorola has leveraged its initial success in MIEL by creating a string of software centres in other regions including Australia, China and now Eastern Europe. Each of these centres has been created on the MIEL model and is seeded by selected employees from MIEL who bring critical best-practice knowledge with them. The offshore software centres have given Motorola a unique competitive edge in software and led to the creation of a separate business division, the Motorola Global Software Group, which focuses on leveraging software as a separate product line. Other firms are watching Motorola's actions carefully as it has shown the way to transform offshore software development from a risky venture into a strategic advantage.

Notes

1 P. Ghemawat, 'The Indian Software Industry at the Millennium', Harvard Business School note # 9–700–036, 31 July 2000.
2 Quoted in Krishna Guha, 'Human Resources in the Software Industry, Survey on South Asian Software', *Financial Times*, July 1 1998.
3 http://www.nasscom.org/articles/sw_export_growth.asp
4 Proposed by the Software Engineering Institute at Carnegie Mellon University, the Capability Maturity Model (CMM) is a widely accepted industry benchmark of an organization's capability in software development. More details are provided in Box 5.1 and at the following WWW site: www.sei.cmu.edu
5 D.C. Johnston, 'IRS Admits its Computers are a Nightmare', *San Francisco Chronicle*, Friday, 31 January 1997.
6 www.cio.com
7 http://www2.cio.com/CIO/expert/viewqanda.cfm?ID=32
8 T. Mayor, 'Hands Across the Waters', *CIO Magazine*, 15 September 2000.
9 *Ibid.*
10 *Ibid.*
11 *Ibid.*
12 K. Ferdows, 'Making the Most of Foreign Factories', *Harvard Business Review*, March–April 1997.
13 This represented around one-sixth of Motorola's total revenues in 1998.
14 The following description of Motorola is taken from the INSEAD case study: S. Dutta and L. Van Wassenhove, 'Motorola India and Software Excellence: Life Beyond CMM Level 5', 1999.
15 By 2000, there were several other software sites in India which were known to be at levels 5 and 4.
16 In a 1998 international comparative benchmark of about 20 different firms which developed software for the Iridium project, MIEL was found to be better by a factor of six in overall productivity and quality as compared to other firms.
17 In late 1998, another Indian company, WIPRO, achieved CMM level 5. Worldwide, of the 20 firms known to be at CMM level 4, six were in India.
18 In areas of MIEL's expertise. Expertise in different domains was divided across the different Motorola Software Centres.

19 The Change Control Board usually comprised the Project Manager, the Software Quality Manager, the Technical Lead (a senior experienced technical member of the project team) and the Test-site Lead (a project team member responsible for implementing the project at the customer's site).

20 *Technology: IT Services*, Goldman Sachs Report, 20 September 2000.

21 *Ibid.*

22 *Ibid.*

6
Resources, Capability and Coordination in Indian Software Firms

Parthasarathi Banerjee

> A software product is of no use in itself, but only when working in conjunction with other complementary products as part of a system. Thus in the simplest case of an isolated personal computer, software products must work with each other; an operating system, specifically, must work with applications and other elements in a hardware platform. In the case of an extended network such as the so-called information highway, the set of related components will be much larger. If these systems, large or small, are to do what we expect of them, then the component parts must be so designed as to inter-operate.
>
> (Richardson, 1997, pp. 10–11)

Introduction

Interoperability of a software service or a product appears to be only one aspect of the larger issue of coordination amongst software-writing firms. The system that a software must be complementary to, is itself evolving, in other words a software introduces a variation and consequently a change in the system Richardson (1997) talked about. A system appearing to be robust often quickly gives in to an emergent standard. Product life-cycles are extremely short, and exchange of technical information between software firms helps this concurrent coordination. Inside a firm, especially if it is large, product planning is coordinated through another kind of information. Beyond these two variants of information, there must be another creative variant of information which sets a disequilibrium in the system interoperability. In order that the new technology that this disequilibrating software is bringing in can become operational in the market, a bevy of firms must commence coordination in anticipation – coordination in expectation (Rosenberg, 1976). This appears as a new mode of coordination. 'Business models' has been a recent attempt in such coordination.

Traditionally, coordination has been achieved through integration of stages of production requiring evolutionary coordination (Richardson, 1998) under one roof of a vertically-organized firm (Perry, 1989). The other mode across several firms producing closely similar or otherwise complementary products has been the network or trust (Richardson, 1960). Market coordination achieved through movements of intermediate goods (Sraffa, 1963) has remained another mode. Recent talk about still another market-based mode – seen in the efficacy of the securities market in gaining corporate control over 'financially weak' firms – has, however, been doubted by Lazonick, who note that 'value-creating investment strategies increasingly require that business organizations exercise control over, rather than be controlled by, the market for corporate control' (1992, p. 53). In fact, it would not perhaps be unreasonable to be circumspect about the spate of acquisitions by large corporate following the meltdown in the dot.coms – an event which definitely smothered the likelihood of technically innovative challenges from the tiny startups. Likely alliances between corporates and fund manager/stock-traders have been known to decide the fate of small beginner firms. These are but some of the modes of coordination.

The history of co-evolution of large corporations and of the corresponding technology (Nakagawa, 1986; Penrose, 2000) shows a pattern marked by products with longer life-cycles. This also allowed time for coordinated firms to invest in new machineries subsequent to any innovation in a large corporation. However, in software there are a few peculiarities. The marginal cost of production of a software is near zero, software service produces an apiece product, investment in fixed capital machineries is comparatively small, and the life-cycle is very short. Moreover, most firms do not, and in fact cannot, undertake proprietary R&D; innovations depend, it appears, on simultaneity of a publicly available advanced knowledge pool and concurrent expectations of several firms on a new set of interoperable products. The historical experience has shown that coordination was achieved most effectively with vertical integration and sometimes even through expansion of the scope. Consequently, competition, including potential competition (Gilbert, 1989), depended on such aspects as entry barriers or competitive advantage. With sunk costs approaching zero and prices moving quickly, established firms can earn profits even following entry (Stiglitz, 1987). The Chicago School variants of efficiency-related competence theorists (Demsetz, 1973; Teece and Pisano, 1994; Leonard-Barton, 1992) emphasizing, in particular, efficiency of resources use (Bromiley and Fleming; Banerjee, 2002), thus appears to be limited by, first, shortening of product life-cycles, and second by the vanishing ability of a manager to procure, control and apportion resources only internally (Banerjee and Richter, 2001). Dynamic competence refers to a firm's internal capability, and refers to apportionment of resources. However, if firms, as in software, have to concurrently expect a set of innovations and consequently if these firms have to concurrently apportion resources not only inside the firm but simultaneously amongst firms, it would not be surprising that the strategy-determined structurally-induced apportionment of resources gives way to a new mode of resources acquisition, use and apportionment.

It is doubtful, then, that software firms will integrate vertically. Further, strategy in tandem with the structure (possibly not vertically-integrated) would have to be adjusted to a concurrent coordination on resources use. In other words, software firms would then appear to be coordinating their resource usages. These two hypotheses are interrelated. We will narrate aspects of the Indian software scenario in order to substantiate these two interrelated claims, but will not attempt to provide any substantive theory. Given this analysis of Indian software certain public policies can be inferred, but we will limit ourselves on the public policy aspects. The resource that software firms primarily use is manpower, and we will discuss the use of manpower by these firms to show that vertical integration is undesirable and inter-firm movements of human resources coordinate the concurrent expectations on a set of systemic innovations. We will proceed by first examining firm differences from data on resources use. Next, having seen that unique resource-use behaviour is uncommon, we proceed to consider the issue of coordination through information acquired from resources. Human elements provide a unique information-generative resource and we will speculate on how the flow of manpower across firms provides a unique opportunity to a manager in accessing and in creatively using the information. The manpower market then provides the desired coordination in expectation of systemic innovation.

The resource-use pattern in Indian software firms

Typically, a software firm invests in land and buildings. Apart from this, investments in hardware and software purchase consumes a very small fraction of the total revenue of a firm. As a result, the ratio between 'investment' to 'total funds' (as defined in the Annual Reports of the firms) can only take up higher values approaching 1 only very occasionally. We studied the value of this ratio for three years 1997, 1998 and 1999 for 29 firms, finding an average value for 1998 of 0.33, and for 1999, 0.04. Distribution of the values was rather skewed (since one or two firms invested heavily in the corresponding year on land, etc.). Often this ratio took values in the range of 0.01. Investments in software were understandably an even lower fraction of total funds. Total funds indicate in accounting terms a fund that is investible. We looked for investments in reported R&D as well, but rarely would a firm invest in R&D. Most firms deployed investibles in money-market instruments. As a result, and in contrast to firms in traditional manufacturing sectors, the firms in software were found to be making little sunk investment. Sunk cost of a software entity is small, and can often be ignored.

However, a software firm spends substantially on its manpower. Although one could argue that 'total personnel costs' (including salaries and wages, bonuses, ESOP, welfare, training, etc) are an investment, we might also compute the ratio of 'total personnel costs' to the sum of 'cost of sales' and 'cost of communications', rather than the more ordinarily used ratios of total personnel costs to either total revenue or total funds. Our proposed ratio can capture several aspects such as the deployment pattern of staff between sales and software writing (since administrative staff are very few), and intensity of use of communications by staff. This ratio, too, was computed for the same three years. Figure 6.1 presents a chart of this ratio for 14 firms for the three years 1997, 1998 and 1999. In none of the reporting companies did this ratio take a value lower than 1, in about five companies it was above or about 2, while for a few the ratio took values of about 7.

Considering that costs of sales include all costs of which salaries constitute a fraction only, we may conclude that the cost of personnel engaged in writing software is substantially higher compared to the cost of personnel engaged in sales. In fact for some firms the former is several times that of the latter. This is indicative of the deployment of the major resource that a software firm procures. Resource deployment inside a firm characterizes the competency of the firm (Leonard-Barton, 1992; Banerjee, 2001a). This resource includes more of internally generated components (Amit and Shoemaker, 1993; Galunic and Rodan, 1998). Market can differentiate across firms based on quantity of resources procured from the market, however, market cannot differentiate between firms based on resources that are generated internally by the competency of the firm. There is a trick here, however. Manpower is such a resource

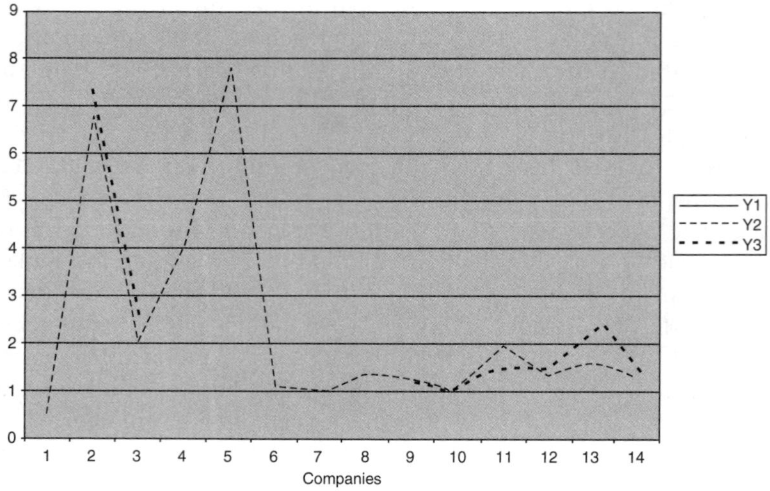

Figure 6.1 Ratio between cost of personnel to cost of sales plus cost of communications: Y1 = 1997, Y2 = 1998, Y2 = 1999

as would always reflect qualities of the firm that it is deserting – therefore, unless the manpower is fresh and inexperienced, manpower acquired from the market would bear characteristics of the knowledge experiences, that is of the competencies of firms being deserted.

Before proceeding further on this issue let us consider another ratio on the relative merit of manpower as resources in a firm's agenda. The ratio between the sum of 'total personnel costs' and 'depreciation' to the 'total revenue' of a firm was explored for 51 firms for which such data were available for the three years 1997, 1998 and 1999. Figure 6.2 shows the chart of this ratio. The numerator includes two types of resources; the first concerns manpower, and the second, depreciation, is more reflective of cash funds made available to the firm. A software firm typically retains a good cash position, in the form of several holdings in money-market instruments, which can act as a buffer to any sudden demand, say on recruiting manpower, emerging out of a sudden new project. The revenues of a firm would often get distributed between these two. As a corollary, we should expect the value of this ratio to hover around the value 1, and that is what we observe in the Figure 6.2. Except for two stray firms, and only for two years, all other values are generally below 1, around 0.75. This indicates that funds and human resources are virtually the only two resources that a firm retains or invests in.

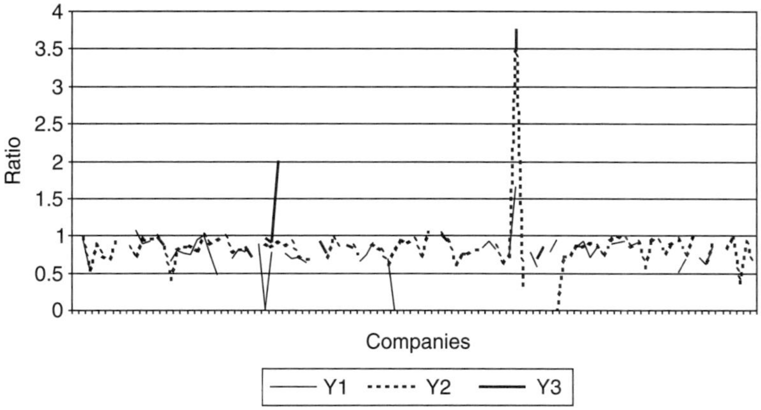

Figure 6.2 Ratio between sum of total personnel cost plus depreciation, to total revenue, for 51 firms: Y1 = 1997, Y2 = 1998, Y2 = 1999

Source: Banerjee, 2003 (with permission of the *Global Business Review*, published by Sage).

With these ratios we are in a position to say that investment in (or cost of) human resources constitutes a major fraction in the investment portfolio of a software firm. Typically such a firm would engage its manpower in software writing more than in, say, sales. Since human resources could be considered as either liquid or plastic, the second ratio indicates a liquidity profile of a firm's resources. The first ratio indicates the use of such a plastic resource by the firm's internal routine which is geared towards achieving a higher competency. A strange point, however, cannot be missed; a core competency is about differentiating a firm (Nelson, 1991) and this is achieved, as per theorists of core competency (Foss, 1999), through an allocation rule and practices of the firm's management. Such a practice is internal. Information on such competencies can be secured by competing firms through two modes: first, through tangible figures reported in public as in Annual Reports; second, through intangible qualities of the manpower who are deserting that firm and who carry information on the knowledge competency of that firm. Manpower performs, then, the twin role of an intermediate good and the carrier of information.

Information flow across firms

Firms in traditional sectors employ economic information on competition, consumers, logistics and even trade-related aspects for their business calculations. Technical information regarding state of the art and benchmarking are used as well for meeting non-information goals of manufacturing or service provisioning. Coordination amongst firms is secured through such information and this information does not in general serve the internal structural purposes of a firm. In the highly innovative information sector, information plays a radically different role. Economic information on competitors, consumers or logistics, and technical information regarding the state of the art, for example, are necessary for firms in this sector although importance of such procured information is markedly diminished. The market for software changes too quickly and information through static channels fail to serve purposes of dynamic coordination that a software firm is always looking for. A software firm here 'generates' much of the information. Generation and management of such information, belonging to both the economic and technical types, demand a different perspective. Managerial attention in software firms gets devoted to this generative aspect.

A software firm locates its information-generating resource in those who are employable and in its employees. An employee becomes a source of both market and technical information only if the employee switches jobs, that is only if he or she becomes a creative carrier of information. Richardson (1960) had talked about these two types of information in the context of a market which returns back to equilibrium quickly through coordinated adjustments brought about by these two types of information. Software, however, challenges this type of coordination. Coordination is sought in expectation of a set of innovations, that is a plethora of innovations which would in the near future provide an alternative coordination. The current ensemble of product producing firms provides a certain coordination achieved through stabilization of exchanges in the above two types of information. The future, to be brought about by another ensemble of products/technologies, would require another mode of information exchange. The interregnum between these two ensembles is punctuated by a new type of information, not considered by Richardson. This we may borrow from Shackle (1972, 1988) and call information-in-expectation. Market and technical information allows a firm to adjust its competency profile and such a profile is decided by the systems and structure of the firm or, in another language, by the resource-usage profile of the firm. Firms in this picture are blocks attempting to adjust internal competencies. In the case that a firm achieves superior technological knowledge or competence, that firm can through mergers and acquisitions enhance both its vertical integration and scope. However, with information-in-expectation, immediate disequilibria is the goal, and firms here cannot generate this information internally. Thus, even if the competency of a certain firm entitles it to go for M&A, it would not gain access to the information-in-expectation that the merged entities could have otherwise provided as independent agencies. The source of such information-in-expectation is entrepreneurial; it is similar to what Hayek (1949) had described in his discovery procedure. The firm as an entrepreneurial agency is a key element which keeps searching for this new type of information.

The human element is the source of this information. Manpower is thus a resource to a software firm in two respects: first, as a resource in the former or ordinary sense; second, as a resource that can provide information-in-expectation. This view of resources stands in contrast to the resources-based view (Barney, 1991) of management, which accepts as resource only such items that remain within the organizational boundary. Information resource as we now observe is dynamic and information-generative, and we treat the job-switching person as the meta-network between firms in competitive expectations. Job-switch helps the strategy coordination of firms, and it sustains development of strategic capability. The other firm is the site for securing information and is also the supplier of information-generative human resources. Job-switching personnel act as an intermediate good that can generate information for innovation-strategy coordination between firms in a highly dynamic and innovative sector of industry.

Resources and coordination: information-in-expectation

Inter-firm competition and cooperation can be meaningful only when firms do differ. In the literature, differences between firms operating in a rather static technological environment have been identified as resulting from practices internal to the firm (Nelson, 1991), from the dynamic competencies (Teece, 1992) or from evolutionary learning (Nelson and Winter, 1982), or else from the mode of resource combinations inside the firm (Peteraf, 1993). These differences have been located in a domain internal to the firm, where these theorists argued, borrowing from Coase (1990), that prices, the only distinct signal from the market, did not operate; firm differences differed on account of non-priced practices. Such a view, however, has been challenged by Richardson (1998) and Foss and Mahnke (2000). Resource allocation inside large corporations does take into account prices in the market. Moreover, human resource, as in our example, when migrating across firm boundaries and whose prices to a procuring firm reflect the market condition inform us similarly that the resources internal to the firm are but reflections of the market. In particular software firms do not impose restrictive conditions on the migration of knowledge workers.

Technical and market information cannot be considered as resource. The resource theorist argues for a distinctive mark which can result from firm internal practices alone. Information available equally to any firm cannot accordingly be considered as resource. The resource theorist Barney (1991) attempts to seek explanations for firm differences from information generated within the firm. Resources are generally defined (Bromiley and Fleming, 2001) as capital, labour, brand, tacit knowledge and so forth and since managerial decisions are assumed to be 'rational' and the market is supposed to be in equilibrium, differences in resource-endowment alone can explain inter-firm differences. Since bounded rational agents are limited in their decision capabilities, prior conditions of resources and of firm history (Amit and Shoemaker, 1993) would appear to influence differences between firms. However, from this perspective information as a resource has to be internal and structural. Strategy and coordination do not feature in this approach and optimality of managerial decisions is strictly limited to markets in equilibrium (Foss, Knudsen and Montgomery, 1995). Intermediate-good-linking firms and their strategies cannot be defined as resource accordingly, and markets with surprises, too, cannot be analysed. Personnel deserting a firm to join another is a strange kind of intermediate good. Software writing involves cognitive capabilities of a team orchestrated together, and little of this cognitive capacity can be captured in the extant objective hardware or processes and most can be located in the human element. Migration of cognitive capacity, which understandably is the principal resource of a firm, is tantamount to migration of an intermediate good. However, in traditional models such as that of Sraffa (1963), an intermediate good links possible vertical relations in an industry or an economy. Human resources in software, however, do not by and large hold such a prospect for vertical linkages. On the contrary, human resources if restricted in job-switching lose their value and will cease to provide the third type of information. As a result, vertical integration which should have followed as per the logic of Coase (1990) fails to appear.

The crucial point then is whether a software firm can have hardware, systems or structures which are not peculiar to any single firm. Such a system or structure would then be generic, that is common in its general form to all firms writing and marketing software. The advantage of such a generic or species-specific structure would be that the flow of resources across firms cannot then significantly harm the competency of the firm. A firm sharing in common with others similar 'innate' or generic structures would not be greatly affected by the flow of resources – this is the view of an evolutionary theorist who criticizes the resource theorist since, according to the latter, resource outflow could ruin the firm competency. Evolutionary theory borrows its paradigm from Chomsky's generative grammar which offers a generic structure and a set of competency-defining generative rules. In evolutionary theory, information procured from the market is specific to the firm; a firm selects and procures information based on its earlier history, its internal structure and routines. Its internal structure is innately given and is therefore common to all. Its strategy and its competency are defined by the rule/routine it follows and the strategy it adopts. The strategy of a firm can adapt itself to both markets in equilibrium and in disequilibria. However, and here lies the crucial difference, an evolutionary firm does not consider information as the differentiator between firms. In other words, evolutionary information is akin to generic information. Information is here specific to the industry, similar to what Richardson (1972) described. Differences between firms, an evolutionary theorist would argue, arise out of firm-internal adaptive or innovative capabilities, and out of routines of information usage that this firm has evolved. An evolutionary routine on information usage is crucial to its theory and such routines inside an organization are internally fixed and internally evolved.

Can a software firm possess such generic systems, structures or hardware/software? Attempts at evolving such generic software or systems have apparently failed in the case of software (Barr and Tessler, 1996; Tessler and Barr, 1997), and human resource practices could also not evolve into innate systems (Baron, Burton and Hannan, 1996). In the absence of any empirical evidence in support of such generic structures in software firms, especially those from India, we can infer that software firms do differ in structures and systems they have evolved into and such systems are not generic. Such systems are peculiar to a firm, a perspective advocated by the resource theorist. It would appear that coordination assumes even greater significance when firms do not share similar generic structures. A firm must coordinate its products, production and resources, especially of information, with its competitors; it must also simultaneously make demands for such information as can differentiate itself from the other firm and it must retain strategic control over the medium of such information. An intermediate good rich in information serves these objectives fairly well because, first, the firm substantially controls its information or cognitive content; second, it can differentiate between firms; and third, it can coordinate functions of firms in both existing and potential ensembles. Human resource alone satisfies all this. Especially in a market full of surprises demand for coordination is raised more and a firm must respond to that. The information sector has such a market and firms here must coordinate strategies through exchanges of such information goods as can generate information on the dynamic frontier of market surprises. An intermediate good that can be generative of information upon demand from the dynamics of the market surprises would serve this purpose.

Numerous small and start-up firms dot the software-writing scenario (Kenney, 2000), and several of them coming out of large organizations as spin-offs often get coordinated by the old boys' network. However, in India, the coordination issue appears complex because most startups are not spin-offs. Coordination pertaining to products and production planning, technological standards or markets-making appear crucial to these small firms. Information required for coordination cannot be procured from the typical markets of products, technology or from fresh graduates. A firm, in order to manage its business and in order to achieve a strategic vantage point of sustained competitive advantage, must access information-generative resources. These two aspects, namely coordination and capability are thus interdependent, and, following Richardson (1972), we argue that the site of firms' capabilities is coordination amongst firms (Foss, 1999; Krafft and Ravix, 2000).

However, in Richardson (1960) a firm uses two types of information only. We have added a third type, following Hayek, and called it information-in-expectation. This third type is not provided by either the systems/structures of market and technology, or by the systems/structures of the firm itself. In other words, this information is not provided by a situation either obtaining in the market or in the firm's organization. It is resident in the personnel crossing firms' boundaries; thus, a person when not related to a firm ceases to provide this type of information. An unattached software professional can provide technical information although he or she cannot provide relational information pertaining to a firm's expectations. The carrier of this information is a person who is aware of the expectation of the firm they are leaving. Moreover, they themselves have been holding this same expectation and, as active agents, they can upon joining another firm who wishes to participate in the innovation ensemble, become generative of the expected information. Such agents therefore act as both carriers/translators and as an active imaginative engine.

Information-in-expectation and coordination

Information-in-expectation is about situations in a firm, although it is not of these situations. It therefore relates to the capability of a firm including the rules/routines of that firm. Further, it relates to the expectations on innovation ensemble that this firm holds. Such an expectation, though related to firm strategy; is distinct from it, and moreover is different from the technological expectations described by Rosenberg (1976). The latter as information can be derived from several public domains, such as scientific literature; a technological expectation is public and therefore a market on technological expectation cannot be created. However, information-in-expectation pertains to firms' internal situations, strategies and expectations. The other firm receiving the experienced job-seeker from the market is in fact in search of the ensemble or the contour of innovation. This receiving firm cannot through technical or market information (for example, on market share, which is often meaningless in software) attempt to coordinate its own expectations with the expectations of the other firm without having accessed this third type of information. Moreover, if these two firms merge together into a single firm, this third type of information will cease to be produced; that is, the possibility of innovation would be lost. We are arguing, therefore, that in order that innovations might appear and in tandem, it is necessary that there remains several firms in expectations of an innovation ensemble and that there gets created a market between these firms exchanging active resources of information-generative personnel. Tacitly we are also arguing that a software firm is founded as much externally as it is internally; the dominant theory of the firm is based on the later. Richardson (1998) attempted a bridge across the firm-capability, an internally constituted element, with problems of coordination. We are extending this to the situation when firms are dynamically innovative, product life-cycles are very short and marginal costs of production are near zero. Firms do invest in relation-specific codes and in relation-specific assets, and thus invest in information-build up as a capability or else strategy of a firm address in a fundamental mode the coordination aspect.

This expectation is also about positioning the firm. The objective of positioning must refer to both current markets and possible future markets, and positioning must refer to other existing or potential firms. Operational aspects reflect current strategic concerns and the structural features of a firm (Chandler, 1990). In software a firm continually keeps shifting its operational features, which are by definition current, to a direction in future. This directionality cannot be secured by a firm which is small or by a firm which, while large, has lost in the process of becoming large the generative sources of innovational information (the latter, it may be recalled, reside in differences between firms). In short, a firm attempts to define and position itself with respect to other firms in both the operational domain and the future domain of the ensemble of innovation. The internal domain of the firm repositions itself in accordance with this strategic and external requirement. This shift in internal domain and in the strategy of the firm then gets reflected in the current operational domain; the operational features of a software firm are highly dynamic.

Operational features, too, get coordinated consequently. This happens because, unlike the case of static coordination, in dynamic coordination operational coordination gets guided by the expectations on the future ensemble of innovation. The firm strategy shifts itself from the previous structure–conduct–performance paradigm (SCPP), to strategy as a design on transforming firm capabilities, systems and structures in tune with the innovative capability that may be required in future. This strategy is undertaken concurrently by firms who all wish to join in the future ensemble (for example, a wireless-driven ensemble). Richardson (1972) argued from the situation of firms coordinating their activities through a dense network of cooperation and affiliation. Such cooperation and affiliation can provide concurrent coordination, hence a predictability of what a competing firm is doing. However, this concurrency refers to a situation we have described as static. Firms would require this concurrency in order to create a market for a new product, or in order to become familiar with what other firms plan to undertake in future or therefore to secure a better predictability of its own supply of resources or of its likely success in marketing. Such a concurrency, too, then directs the operational aspects, but from a current perspective only – 'the need for concurrent coordination ... may reside in their *similarity*, in the sense that they require, ... the same capabilities for their successful undertaking' (Richardson, 1998, p. 5, italics in original). This 'similarity' can be compared to what we have described above as generic structures.

Operational and even strategic issues tend to be 'similar', or tend to acquire same generic structure provided that the context is technologically static. However, for software, the context is different calling for, first, coordination of expectations, and then, second, in order to be similarly capable in future, it also calls for coordination of operational aspects. This later coordination is guided by the former, and therefore personnel flows which provide information-in-expectation and which coordinate future expectations also coordinate the operational dynamic features. Such a flow of this strange intermediate good achieves a cognitive similarity, what Nooteboom (1999) calls similarity as a converse of 'cognitive distance', a measure of which is indeed necessary in order to forge alliance or concurrency. Predictability and the pay-offs of the information secured (Monteverde, 1995; Constant, Sproull and Kiesler, 1996) can be ascertained only in future. Future uncertainty becomes reduced by coordination of strategies and dynamic operational features. Exchanges of intermediate goods that can also generate information, and that can reduce cognitive distance in a possible world of market surprises, must be accommodated in the framework. This good we have located in information-personnel switching jobs.

Job-switching in Indian software

A very high rate of manpower turnover is characteristic of the software industry in several locations, notably in Silicon Valley and in India. Biotechnology, although another frontrunner, does not exhibit similar features. We would argue that in biotechnology a firm has technological expectations (Rosenberg, 1976), but it does not have expectations on the ensemble of innovation. Typically software has evidenced a plethora of products, of which there are several sets or ensembles, and it has also evidenced even more numerous processes. This has not been evidenced in biotechnology for drugs and pharmaceuticals, or biotechnology for agriculture – in both of which patenting by large corporations appears to have severely restricted the spate of innovations. Moreover, a software product requires much less capital to launch than would be required in the case of biotechnology. In short, the innovation profile in software has exhibited certain unique features. We have been describing a possible explanation for such a sustained high rate of innovation in hordes or ensembles, and the explanatory key to our understanding is job-switching personnel.

Job-switching across software firms ranged from over 40 per cent (TCS Report, 1991) in the early 1990s to at least 25 per cent around the late 1990s in India (Banerjee, 2001b). Surprisingly, software firms did not appear to have suffered from this; Indian information-sector firms experienced an average of above 50 per cent growth throughout this decade, and software turned out to be a key export earner. The number of firms increased several fold, and several global corporations opened up their own units during this period in India. Manpower output with a software background also grew several fold. Inter-firm job-switching never came down during this entire decade, and a large number of personnel migrated to other global locations. All this indicates a market for software manpower which had never existed before 1990 had come into existence in parallel with the emergence of the firms. This market, however, as we will soon observe, transacted mostly in experienced personnel.

Data on job switches were collated into a database created from the open advertisements that appeared during September 1998 to March 2000 in a popular Indian business magazine for the information industry. Data on a total of 1615 advertisements for more than 10 000 jobs were covered. It should be noted that the total number of job switches is much higher than the advertisements indicate since Indian firms induct a substantial fraction of their employees from the 'old-boys' network, and fresh employees are often recruited by large firms directly from the

Table 6.1 Skill categories required by software firms

Requirement skill
CAD/CAM/GIS
Client/Server
Client/Server, Web designing/development
Client/Server, Operating systems
Domain knowledge
ERP/CRM tools
Hardware
IBM-mainframe softwares
Languages
Language, Client/Server
Language, Client/Server, W/D
Language, Web designing/development
Not mentioned
Operating systems
Operating systems, Client/Server
Operating systems, Language
Other
Sales
Software testing tools
Systems administration, Language
Systems software
Telecom
Telecom, Client/Server
Tools and softwares for DBA
Web administration, Client/Server
Web administration
Web administration, Operating Systems
Web designing/development

campus. Freshers without experience cannot provide the third type of information, although those who are recruited from the old-boys' network and are from other firms do indeed provide information-in-expectation. Advertisements probe into the scope of coordination of expectations that the market can provide. These advertisements were in prose, and thus the desired types of experiences, qualifications or the advertised job positions had to be classified and clubbed together under several types as shown in Table 6.1 and 6.2 (Banerjee, 2002). Only the years of experience required were given in exact numbers. Table 6.1 presents skill categories under 28 groups with a large degree of overlap. Other categories could have been created, but without causing significant changes in our understanding. Similarly, desired areas of experiences were categorized under several headings (Table 6.2), again with large overlaps.

Figure 6.3 shows a profile of experience against skills required from these advertisements. It may be seen that for the same skill, several areas of work experience count or, conversely, for the experience, several types of skills are counted. It was observed (Banerjee, 2002) that skills requirements varied widely for the same type of job. All this is indicative of the

Table 6.2 Area of Experience demanded by software firms

Area of experience
Database administration
Designing tools
Education
Electronic design automation
Embedded systems development
ERP
Hardware design
Mainframe
Not mentioned
Other
Project leader
Project management
Quality assurance
Sales/marketing
Software development
Solution architect
Systems administration
Telecom
Web administration
Web design/development

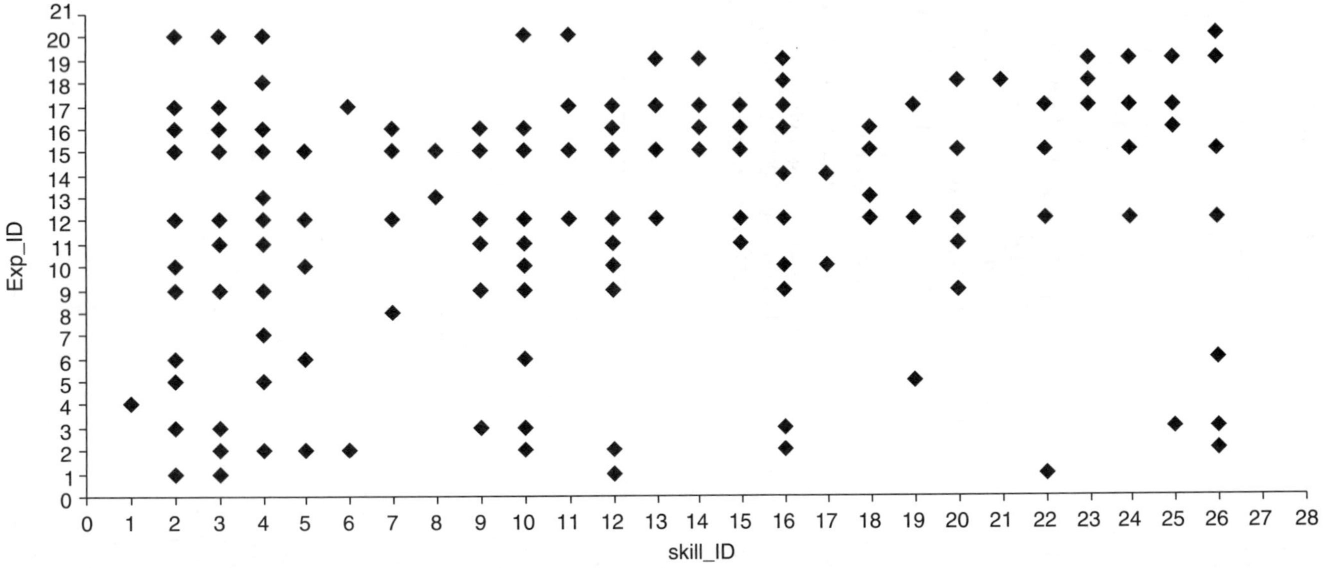

Figure 6.3 Skills by experience demands of firms as reflected in advertisements

Source: Banerjee, 2002 (with permission of *Global Business Review*, published by Sage).

fact that firms follow peculiarly eccentric definitions on areas of experience, skills that are on offer by the employee, and skills that the advertised job would require. Firms appear to follow, as revealed through interviews with firm managers, their own judgment on the types of skills, experience and information-in-expectation that the prospective employee might be able to offer. Further, the nature of job-offers and the years of desired experience did not show any specific pattern (Banerjee, 2001c). However, across almost all kinds of job-offers, persons with about up to eight years of experience appear to be in demand. Similarly, persons with up to eight years of experience in any specific areas are equally attractive in the job market. Contrary to popular belief that in software experienced personnel tend to stick to the old firm, our observation shows that experience makes personnel even more attractive to the prospective employer.

Firms demand persons with experience. However such demands are not articulated as clear and distinct signals to the market – signals are often elusive and much of the transactions between prospective employers and employees take place at the interview board. Often, advertisements indicate little more than the fact that there exist job opportunities. The exact number of jobs offered by a firm is not disclosed in an advertisement, hence the total number of jobs for experienced persons are not known. However, it is strongly indicated that experienced persons are in greater demand by firms in software. Job signals fail to articulate on exact correlation between types of jobs and their respective demands, or their respective potential applications. This implies that the potential job-seeker remains dependent on the 'old-boys' networks or on professional networks for such extra information. Further, an employee would then like to carry more information on the firm he or she wishes to desert, or on areas of technologies that are related to their current jobs. The former implies that an employee has an incentive to become a better carrier of information-in-expectation; and the latter that she has an incentive to acquire more knowledge on the potential innovation ensemble and its technologies. Both these therefore assist information passage across firms, and such information passages are coordinating in nature.

It is not strange, therefore, that the high manpower turnover has not disturbed the growth in revenue, the market or in technology or innovative products. Instead, all the firms experienced growth in all these areas. The average sizes of all the firms also grew during this period. Manpower turnover does affect firms in old types of industry, however, who do not require coordination of their expectations on innovation. The knowledge management practices of such firms, as typical of a Japanese firm, emphasize tacit knowledge and knowledge located in job/tasks that is non-transferable (Nonaka and Takeuchi, 1995). A typical firm might then talk about lifetime employment and the firm's personnel practice would ensure restrictions on personnel mobility. Often such restrictions appear in the form of legally-binding clauses restricting floor-crossing through threats of penalty. Such penalties are imposed mostly on those employees who recruited as freshers and having spent substantial years with a particular job/task of a certain firm, acquire knowledge which has little market outside the firm, and conversely the firm, too, cannot replace that person with an experienced hand secured from the market. Such a picture is missing in software. Most software firms hire freshmen and put them through rigorous induction training followed by on-the-job training. A fairly large number of persons switch jobs immediately following such trainings, and this practice of job-switch is generally maintained by the person. Such trainings, however, are not given to new recruits with experience. Rarely would a firm apply strict legal rules in employment contracts, and there is no penalty ordinarily imposed on job quitting. Software firms often retain a pool of extra manpower resembling the inventory of a manufacturing firm, and such a pool is either put on training in areas that the firm expect to develop, or is assigned to a developmental project/product. Upon the sudden start-up of a new project, this pool is disbanded and works on the new project.

Concluding remarks

Job-switching has not compelled software firms to vertical integration. Very few such firms have even gone in for mergers in the direction of enhancing their business scope. Job-switching appeared not to have affected the personnel policies of these firms who could in this interregnum increase their business immensely. If manpower were a resource, as per the resource-theorist perspective, a loss of experienced persons should have proved if not fatal at least emaciating. An experienced person is a valued resource, in particular for software firms. Such firms employ little of other resources and most of their strategic and operational activities are designed around manpower which is thus very critical to such a firm. Yet loss of such an important 'resource' is strangely accommodated in the strategies of software firms. It seems that job-switching has become part of the managerial strategy. Recruitment of personnel is often not through campus interviews for freshers or through open advertisements in the business press; a fairly high percentage of personnel are recruited through personal networks (Banerjee, 2001b). Information in this personal network, however, mostly remains territorially limited. On the other hand, switching jobs across distant cities or countries often depends on open advertisements. Recruitment firms act as market intermediaries who through their knowledge of the market and their databases act as matchmakers.

The position of the resource theorist is thus suspect, and equally suspect is evolutionary theory. To those theorists manpower is also important, in that they hypothesize certain generic structures are common to all firms, especially they are from the same industry. The generic structure alone enables emergence of competency exhibited in routines and rules. However, we observed that software firms use market signals that are elusive and that hide any possible common generic structure. If routines and rules were determining competency in software, as they appeared to have achieved in firms not in expectation of innovation ensembles, a person as a resource would become affixed to a generic structure and the loss of that person would have proved detrimental to the growth of firm competency. Instead, our observations indicate that personnel movement has not affected firm capability; instead, capability has increased. Competency of a software firm therefore does not seem to rest on the assumptions of evolutionary theorists.

Theories of the neo-Austrians, Foss (for example 1997) and that of Richardson (1998) are the closest to what we have argued. We borrowed from Richardson the crucial importance of coordination amongst firms, and an interesting insight is taken from Shackle (1972, 1988) in our discussion of expectations where we have tried to emphasize the pivotal role that they play in innovations in software. These innovations are rarely of a singular path-breaking type – a type that has been much discussed in the history of innovations in chemicals, semiconductors and pharmaceuticals instead, innovations in software appear in droves, in what we have called innovation ensembles. Expectations on innovation in one firm therefore have to be coordinated with expectations by other firms desirous of joining the ensemble. These firms require to exchange, in addition to the two types of information described by Richardson, another new type which we have called information-in-expectation. Richardson's understanding situated a firm on sites both internal and external to the firm, thus linking capability to coordination. Our extension links capability to coordination in future – such a possibility of coordination then drives the firm's current strategies and operational tactics. Our argument is that a software firm's strategy is twofold: first, to be with the drove by way of coordinating expectations; second, to differentiate oneself from the herd of ensemble participants. This explains how dynamic capability is related to strategy, and this process view of strategy is made possible through the passage of the flux of human resources across firm boundaries.

The two types of information described by Richardson (1972) are available to stationery aspects of technology and the market. Information-in-expectation must, however, link a firm's resources-positioning or its dynamic capabilities with the expectations of others. Such information cannot be accessed from Richardson's static sources. Software's only resource appears to be its knowledge workers, and information on dynamic capability resides in this manpower. It is only in exchanging such information resources that coordination in the future can be secured. Coordination transactions across competing firms is thus sustained by a continuous flow of personnel across the porous boundaries of a firm. Our observation is that, excepting very senior people, most of the experienced personnel keep on changing jobs. An experienced person alone has knowledge about the dynamics of business, business expectations, and expected ensembles of innovation. Such a person deserts a project often midway, and gets transacted, as it were, as an intermediate good. A software intermediate good in the form of experienced job-switching personnel links expectations and dynamic capabilities of separate software-writing centres. A vertical integration or even a horizontal merger of separate software firms into a singular firm would force devaluation of the information-in-expectation (such tasks are in this case carried out by a governance-providing hierarchy), and hence human resources would lose value, consequent to which innovations would also fall.

There is thus an uncertainty arising from the flux of this intermediate good apart from the ordinary uncertainty associated with the functioning of the market. The payoff from the new information has to be balanced with the increase in uncertainty. A firm's strategic management of information appears then as a balancing trick between these two poles. 'Each of us can endeavor to reduce uncertainty about what other people will do by exchanging undertakings with them, but only at the price of reducing also the freedom of maneuver with which we hope to cope with uncertainty of a more general kind' (Richardson, 1998, p. 9). Maneuverability is least affected when the information system of the firm and the firm strategy are based on a flow of personnel representing an information-generative intermediate good. In this novel arrangement, mediated through the personnel factor market, information firms coordinate their production of information goods and technology/ services through a new strategic understanding. This strategic understanding is based on the information-generational character of the creative resources that are flowing across the boundaries of the firms. Such movements balance the increased uncertainty with the new sources of information. Hence, the strategy in such a software firm is crucially dependent on the aspects of information generation required for reducing uncertainty as also for enhancing capability. This is made possible by the flow of resources personnel.

References

Amit, R. and Shoemaker, P. (1993) 'Strategic Assets and Organizational Rent', *Strategic Management Journal*, vol. 14, pp. 33–46.

Banerjee, P. (2001a) 'Resource Dependence and Core Competence: Insights from Indian Software Firms', *Technovation* (forthcoming).

Banerjee, P. (2001b) 'Some Indicators of Dynamic Technological Competencies: Understanding of Indian Software Managers', *Technovation* (forthcoming).

Banerjee, P. (2001c) 'Strategic Management of Information in Indian Information Sector Firms through Job-Switches', *International Journal of Information Management* (forthcoming).

Banerjee, P. (2002) 'Resources Strategy and Signals to Resources Market in Indian Software', *Global Business Review*, vol. 4(1), January–June.

Banerjee, P. and Richter, F.J. (2001) 'Social Management: Situating imagination, Concept, Cooperation and Intangible Assets in the Knowledge Business', in P. Banerjee and F.J. Richter (eds), *Intangibles in Competition and Cooperation: Euro-Asian Perspectives*. Houndmills: Palgrave Macmillan, pp. 289–338.

Barney, J.B. (1991) 'Firm Resources and Sustained Competitive Advantage', *Journal of Management*, vol. 17, pp. 99–120.

Baron, J.N., Burton, M.D. and Hannan, M.T. (1996) 'The Road Taken: Origins and Evolution of Employment Systems in Emerging Company', *Industrial and Corporate Change*, vol. 5(2), pp. 239–75.

Barr, A. and Tessler, S. (1996) 'The Globalization of Software R&D: The Search for Talent', Stanford Computer Industry Project, Stanford. See http://www-scip.stanford.edu/scip/

Bromiley, P. and Fleming, L. (2001) 'The Resource Based View of Strategy: An Evolutionist's Critique', in M. Augier and March, J.G. (eds), *The Economics of Choice, Change, and Organizations: Essays in Memory of Richard M. Cyert*. Cheltenham: Edward Elgar.

Chandler, A. (1990) *Scale and Scope: The Dynamics of Industrial Capitalism*. Cambridge, Mass.: Belknap Press of Harvard University.

Coase, R. (1990) 'The Nature of the Firm', in O.E. Williamson (ed.), *Industrial Organization*. Aldershot: Edward Elgar, pp. 3–22.

Cohen, S.S. (1998) 'Social Capital and Capital Gains, or Virtual Bowling in Silicon Valley', BRIE Working Paper no. 132, September, Berkeley, California.

Constant, D., Sproull, L. and Kiesler, S. (1996) 'The Kindness of Strangers: The Usefulness of Electronic Weak Ties for Technical Advice', *Organization Science*, vol. 7(2), March–April.

Cyert, R.M. and March, J.G. (1963) *A Behavioural Theory of the Firm*. Englewood Cliffs, N.J.: Prentice-Hall.

Demsetz, H. (1973) 'Industry Structure, Market Rivalry and Public Policy', *Journal of Law and Economics*, vol. 16, pp. 1–9.

Foss, N.J. (1999) 'Capabilities, Confusion, and the Costs of Coordination: On Some Problems in Recent Research on Inter-Firm Relations', DRUID Working Paper no. 99–7.

Foss, N., Knudsen, C. and Montgomery, C.A. (1995) 'An Exploration of Common Ground: Integrating Evolutionary and Strategic Theories of the Firm', in C.A. Montgomery (ed.), *Resource-based and Evolutionary Theories of the Firm: Towards a Synthesis*. Norwell, Mass.: Kluwer.

Foss, N. and Mahnke, V. (2000) 'Strategy Research and the Market Process Perspective', in J. Krafft (ed.), *The Process of Competition*. Cheltenham: Edward Elgar, pp. 117–42.

Galunic, D.C. and Rodan, S. (1998) 'Resource Recombinations in the Firm: Knowledge Structures and the Potential Schumpeterian Innovation', *Strategic Management Journal*, vol. 19, pp. 1193–201.

Gilbert, R.J. (1989) 'The Role of Potential Competition in Industrial Organization', *Journal of Economic Perspectives*, vol. 3(3), Summer, pp. 107–27.

Hayek, F.A. (1949) *Individualism and Economic Order*. London: Routledge & Kegan Paul.

Kenney, M. (2000) 'A Note on the Comparison between Cambridge, England and Silicon Valley', BRIE Research Note #6, July, Berkeley, California.

Krafft, J. and Ravix, J.L. (2000) 'Competition and Industrial Coordination', in J. Krafft (ed.), *The Process of Competition*. Cheltenham: Edward Elgar.

Lazonick, W. (1992) 'Controlling the Market for Corporate Control: The Historical Significance of Managerial Capitalism', in F.M. Scherer and M. Perlman (eds), *Entrepreneurship, Technological Innovation, and Economic Growth*. Ann Arbor: The University of Michigan Press, pp. 153–200.

Leonard-Barton, D. (1992) 'Core Capabilities and Core Rigidities: A Paradox in Managing New Product Development', *Strategic Management Journal*, vol. 13, pp. 111–25.

Monteverde, K. (1995) 'Technical Dialog as an Incentive for Vertical Integration', *Management Science*, vol. 41, pp. 1624–38.

Nakagawa, K. (1986) 'Business Strategy and Industrial Structure in Pre-World-War-II Japan', in K.Nakagawa (ed.), *Strategy and Structure of Big Business*, Proceedings of the First Fuji Conference, Tokyo: University of Tokyo Press, pp. 3–38.

Nelson, R.R. (1991) 'Why Firms Differ and How Does it Matter?', *Strategic Management Journal*, vol. 12, pp. 61–74.

Nelson, R.R. and Winter, S.G. (1982) *An Evolutionary Theory of Economic Change*. Cambridge, Mass.: The Belknap Press.

Nonaka, I. and Takeuchi, H. (1995) *The Knowledge-Creating Company*. Oxford: Oxford University Press.

Nooteboom, B. (1999) *Inter-Firm Alliances: Analysis and Design*. London: Routledge.

Penrose, E.T. (2000) 'Vertical Integration with Joint Control of Raw-Material Production: Crude Oil in the Middle East', in P. Stevens (ed.), *The Economics of Energy*, Vol. II. Cheltenham, UK: Elgar Reference Collection, pp. 230–46.

Perry, M.K. (1989) 'Vertical Integration: Determinants and Effects', in R. Schmalensee and R.D. Willig (eds), *Handbook of Industrial Organization*, vol. 1. Amsterdam: North Holland, pp. 185–255.

Peteraf, M.A. (1993) 'The cornerstones of Competitive Advantage: A Resource-Based View', *Strategic Management Journal*, vol. 14(2), pp. 179–91.
Richardson, G.B. (1960) *Information and Investment*. London: Oxford University Press.
Richardson, G.B. (1972) 'The Organisation of Industry', *Economic Journal*, vol. 82, pp. 883–96.
Richardson, G.B. (1997) 'Economic Analysis, Public Policy and the Software Industry', DRUID Working Paper no. 97-4.
Richardson, G.B. (1998) 'Production, Planning and Prices', DRUID Working Paper no. 98-27.
Rosenberg, N. (1976) 'On Technological Expectations', *The Economic Journal*, vol. 86, pp. 523–35.
Shackle, G.L.S. (1972) *Expectation in Economics*. Cambridge: Cambridge University Press.
Shackle, G.L.S. (1988) *Business, Time and Thought*, ed. by S.F. Frowen. London: Palgrave Macmillan.
Sraffa, P. (1963) *Production of Commodities by Means of Commodities*. Bombay: Vora & Co.
TCS (Tata Consultancy Services) (1991) *Study on Employment Characteristics of Software Manpower*. New Delhi: Department of Science and Technology.
Teece, D.J. (1992) 'Strategies for Capturing the Financial Benefits from Technological Innovation', in N. Rosenberg, R. Landau and D.C. Mowery (eds), *Technology and the Wealth of Nations*. Stanford, Cal.: Stanford University Press, pp. 175–206.
Teece, D.J. and Pisano, G. (1994) 'The Dynamic Capabilities of Firms: An Introduction', *Industrial and Corporate Change*, vol. 3(3), pp. 537–56.
Tessler, S. and Barr, A. (1997) 'Software R&D Strategies of Developing Countries', Stanford Computer Industry Project, Stanford. See http://www-scip.stanford.edu/scip/

7

Plant Breeding in an Era of Privatization: Reflections on Transformations in the Indian Seed Industry

*Dwijen Rangnekar**

Introduction

In policy terms, the 1980s formed a watershed in the transformation of the Indian seed industry. In 1983 a policy to release publicly-bred varieties to the private sector for multiplication was introduced, which was followed by a marked relaxation of industrial licensing regulations in 1987. Eventually, in 1988, a new seed policy was announced. The changed regulatory environment marks a significant departure by allowing large private sector companies to enter the industry across a range of crops, while also relaxing the constraints on seed imports (cf. Table 7.1). These changes were accompanied, particularly towards the end of the decade, by deliberations on the appropriateness of plant breeders' rights (PBRs) for India (Anon., 1990; Wood and Bombin, 1990). A number of draft legislations have been circulated in India (Dhar and Chaturvedi, 1998; Dhar, Pandey and Chaturvedi, 1995; Rangnekar, 1998), and the latest draft (1999) was passed as law – Plant Variety Protection and Farmers' Rights Act 2001.

The national debate has been influenced by international developments such as advice from multilateral funding agencies like the World Bank and obligations arising from multilateral trade negotiations like the TRIPs Agreement. Interestingly, even while the appropriateness of PBRs was debated in India prior to the completion of the Uruguay Round negotiations, it is the TRIPs-obligation that motivates the law providing

* The comments of Geertrui van Overwalle, Biswajit Dhar, Bharat Ramaswami, C.S. Srinivasan and C. Niranjan Rao on earlier drafts of this chapter are gratefully acknowledged. The usual disclaimer applies.

Table 7.1 Policy developments in the Indian seed industry

Year	Policy	Key elements
1983	Seed Control Order	Places seeds of food crops, fruits, vegetables, cattle fodder and jute on the essential commodities list
1987	Revision to Industries Development and Regulation Act (1956)	Appendix I of the 1956 Act is revised to allow entry of MRTP and FERA companies into seed production of high-yielding varieties and hybrids
1988	New Policy on Seed Development	Relaxing seed import control across all species: (a) seeds of coarse cereals, pulses and oilseed imports allowed by (domestic) companies that have collaborated with foreign companies, with requirement of disclosure of parent lines within two years of commencement of imports; (b) seeds of vegetables, flowers and fruits placed under Open General Licence; (c) wheat and paddy seeds remain excluded from this regulatory change. There are also limited quarantine requirements
1989	Plants, Fruits and Seeds Order	Imports of seeds and planting material of vegetables, flowers and ornamental plants permitted without license
1991	(New) Industrial Policy	Select components of the seed sector – certified high-yielding, hybrid seeds – placed in list of industries for automatic approval of foreign technology agreements and foreign equity

for PBRs according to the Preamble of the Protection of Plant Varieties and Farmers' Rights Act 2001:

> And whereas India, having ratified the said Final Act, should, *inter alia*, make provisions for giving effect to sub-paragraph (b) of paragraph 3 of article 27 in Part II of the Agreement on Trade Related Aspects of Intellectual Property Rights relating to the protection of plant varieties; And whereas, to give effect to aforesaid sub-paragraph (b) of paragraph 3 of article 27, it is considered necessary to undertake measures for the protection of the rights of the plant breeder and farmers to encourage the development of new varieties of plants:

Private sector investments in agriculture research are estimated at 15 per cent of the total and concentrated on few crops (Ravishankar and Archak, 2000), although this will surely change in the future. The increasing presence of the private sector in plant breeding raises questions concerning the future orientation of public plant breeding. On the one hand it provides an opportunity for the public sector to reallocate its scarce resources on crops and regions that remain neglected. However, on the other hand, there are concerns about the monopoly control being exercised over the 'seed' suggesting the need for public investments to maintain key resources in the public domain. This chapter examines issues concerning the mandate for public plant breeding in an era of advancing privatization of plant breeding. The analysis begins by presenting a conceptualization of the 'seed', where two distinct properties (genetic information and physical attributes) are identified and the centrality of 'seed' in the ongoing transformation of agriculture is recognized. This is followed by a brief overview of the policy literature on the appropriate roles for public and private plant breeding. It is with this background that a critical assessment of recent contributions on the liberalization of the seed industry in India is undertaken. The final section presents two broad rationales for developing and maintaining a public plant breeding programme in India.

Seeds as delivery mechanisms of technical change[1]

At issue is the historical struggle to exercise control over 'seed'. Contemporary legal developments aimed at providing deeper and tighter rights have an origin traceable to periods of colonial occupation (Fowler, 1994). During the colonial period control was exercised by acquiring physical possession of key plants and establishing botanical gardens (Brockway, 1979) and the strategic movement of key plants across the Empire – a process that has been christened a 'botanical chess game' (Mooney, 1983). Analysing the changing dimensions of control exercised over 'seeds' and understanding the pressures to dilute the presence of public agricultural research warrants a proper conceptualization of the properties of 'seeds' and the seed's role in transforming agriculture.

It is well-recognized that 'seeds' substantially determine the production frontier by establishing the upper limit to output and the productivity of all other inputs (Cromwell, 1990; Jaffee and Srivastava, 1992, pp. 1–2). However, this characterization of the 'seed' mystifies two separable properties inherent in 'seeds': (a) genetic information, that is *software*, that is the result of breeding programmes, and (b) physical properties, that is *diskette-like* features, that are determined by the seed production process (Lewontin and Berlan, 1990). Each of these two properties are the result of distinct agents and occur at different stages in the production of 'seeds' and are governed by different regulations.

Identifying these properties allows attention to be focused on efforts directed at manipulating the genetic software of plants. The natural and empirical character of agriculture with its peculiar constraints of biological time, places barriers to the entry of capital into agriculture (Bye and Fonte, 1994). These barriers are not immutable and fixed, rather they act as focusing devices for science to develop new techno-economic solutions to circumvent or relax the constraint.[2] For example, consider the biological time required for completing a cycle of plant growth. Here, glasshouses and growth rooms have more than halved the required growth cycle, thus significantly quickening the process of selecting new cultivars (Bingham and Austin, 1993, p. 183). It is through the discrete and discontinuous solution to these barriers that agriculture is integrated into the wider spheres of accumulation in the economy, while also creating a market for industrially-produced inputs into agriculture (Goodman and Redclift, 1991; Goodman, Sorj and Wilkinson, 1987). For example, mechanization replaces animal and human labour in specific chores related to farming without appropriating all labour activities, while also creating a market for mechanical devices.

Manipulation of genetic software occupies a particular primacy in the transformation of agriculture. For example, the introduction of inorganic chemicals relaxes the constraint of renewing soil fertility after each cycle of cultivation. However, transformations at the varietal level – in the plant's genetic software – are necessary to enable the plant to withstand and effectively utilize applications of inorganic chemicals. The inclusion of inorganic chemicals in farming practices requires the development of varieties that are selected to respond to the application of these inputs. To be clear, it is the breeding of short and stiff-stalked plants that enables the effective introduction of inorganic chemicals into agricultural practices. Even while the availability of inorganic chemicals might provide the *focus* for breeders to select for shorter, stiffer stalks, it is the breeding achievement – manipulation of embedded genetic software – that is the crucial contingency in transforming agricultural practices.[3] This importance of the breeding effort in the transformation of agriculture is on account of the malleability of plants. In this sense, machines do not have the selectivity and flexibility to deal with the variability of plants. Consequently, efforts to transform agriculture require a movement 'back to the plant, and indeed, even back to the seed from which the plant comes ... Machines are not made to harvest crops; in reality crops must be designed to be harvested by machines' (Webb and Bruce, 1968, p. 126).

It is also recognized that transformations effected at the level of the 'seed' enable the interlinking of markets providing inputs for agriculture. Neither the specific inputs nor particular types of varieties (e.g. dwarfs) can independently proliferate as the dominant mode of agriculture. Rather, it is a technology package consisting of specific varieties, compatible inputs and defined agronomic practices. However, in the era of biotechnology, evidence of the interlinking of agricultural product markets via transformation achieved at the level of the seed is more transparent. Take for example the breeding focus on herbicide tolerance: adoption of a particular variety ties the farmer to a specific brand-name herbicide. Not surprisingly, herbicide tolerance is the dominant trait being tested across all genetically modified crops: the percentage of transgenic crops with this single trait increased from 23 per cent (0.6 ha) to 54 per cent (6.9 ha) of the total global area under transgenic crops between 1996–1997 (James, 1998).

From the above exposition we draw two implications concerning the importance of 'seeds'. First, manipulations in genetic software occupies a particular primacy in delivering technical change to agriculture. And, following from this is the result that markets for industrially-produced agricultural inputs are being created and interlinked by the breeding effort. Consequently, exercising influence over the plant-breeding research agenda allows a degree of control over the diverse industries feeding inputs into agriculture. Using this conceptual background we examine questions concerning changes in the regime of intellectual property rights (IPRs) and the efforts to redefine the mandate of public agriculture research in India.

The policy dimension

The transparent primacy accorded to the international obligation in the Preamble of the Law (see Introduction) is important to keep in mind when assessing the issue of PBRs in India. The discussion here briefly reviews different perceptions on the appropriateness of the introduction of PBRs in a developing country. As will be evident, the dominant view favours the introduction of PBRs and the simultaneous withdrawal of the state from some/all plant breeding activities.

Accounts of seed-system development reveal a linear trajectory in their assessment of transformations in the seed industry (e.g. Jaffee and Srivastava, 1992; Pray and Ramaswami, 1991). In the 'early' stages of seed-system development farmers tend to be the primary and dominant source for varietal improvement. The 'middle' stages are characterized by the dominating presence of public sector breeding activities and the mixed presence of private–public agencies in downstream seed-system activities. At this stage, as far as varietal development is concerned, private sector investments tend to focus on high-value/low-volume commercial crops and hybrids. The final stage of a 'mature' seed system is distinguished by the presence of private sector companies in varietal development. This classification system raises questions on the appropriate indicator used to differentiate seed systems. Apparent in the scheme is a focus on varietal development, but even in a mature system like that of the USA the parentage of varieties suggests a dominating presence of the public sector: 50 per cent of the wheat and soybean varieties, 90 per cent of barley and dry bean varieties and 95 per cent of rice varieties consist of publicly-bred varieties (Knudson, 1990), and 72 per cent of hybrid corn varieties had at least one parent of public sector origin (Butler and Marion, 1985).

Following the linear classification, the literature proposes normative prescriptions for the *appropriate* roles of the public and private sector (Jaffee and Srivastava, 1992, 1994) (cf. Table 7.2). The prescriptions tend to follow stereotypical characterization, wherein the 'private sector is

Table 7.2 Techno-economic factors and appropriate roles for the public and private sectors

Activity	**Scope for appropriability**	**Externalities**	**Scale economies**	**Private sector incentives**	**Public sector provision**
Varietal development					
• self-pollinated	Low	X	X	Low	High
• hybrids	High	X	X	High	Variable*
Varietal maintenance	Medium	X	X	Medium	High
Seed production					
• self-pollinated	Low	–	–	Medium	Variable**
• hybrids	High	–	X	High	Low
Seed certification	Low	X	X	Medium	High
Seed marketing					
• storage	Variable	–	X	Variable	Variable
• promotion	Medium		–	High	Medium
• distribution	Medium	X	–	High	Low

Notes: * Depending on the level of scientific development, early stages suitable for public sector involvement. ** High justification for public sector involvement in foundation seed production and at early stages of seed industry development. X indicates the presence of the relevant property.
Source: Jaffee and Srivastava (1992).

normally flexible, responsive to changing requirements' (Gregg, Delouche and Bunch, 1980, p. 24). In contrast, the public sector is invariably considered as being 'inefficient or ineffective' (Jaffee and Srivastava, 1992, 1994) and under significant budgetary pressure (Morris, Singh and Pal, 1998; Mruthyunjaya and Ranjitha, 1998; Tripp and Pal, 2001).

There also exist contributions that adopt a more cautious approach. For example, Barton (1982) remains doubtful about the appropriateness of PBRs for developing countries because the socioeconomic tradeoffs are quite different. Others agree that there is little evidence of the direct benefits of introducing PBRs in developing countries (Lesser, 1991). The caution, it is often suggested, testifies to the fact that adverse implications of Intellectual Property Rights (IPRs) tend to be easily foreseeable in contrast to possible benefits like increased investments and food security (Cohen, 1999). Policy development requires a detailed empirical assessment of the wider effects of adopting PBRs (Godden, 1984). To be fair, this requirement appears, though somewhat fleetingly, in the seed-systems literature reviewed above. For example, the policy advice is to gauge 'the incentives for private sector involvement and the justification, if any, for public sector intervention' (Jaffee and Srivastava, 1994, p. 103). More recently, given the obligation under the TRIPs Agreement, IPGRI (1999) has developed a policy document identifying key questions for legislators in developing countries.[4]

This brief review demonstrates that different views exist on how to demarcate the *proper* areas of activity for the public and the private sector, reflecting divergent perceptions on the appropriate balance of interests between social groups engaged in agriculture. There is little doubt that the public sector is under increasing pressure to reorient its presence in agriculture and, following the introduction of PBRs, to withdraw from *near-market* activities, viz. seed production and distribution and varietal development.[5]

Seed industry liberalization in India

The initial reaction to the 1980s liberalization have been predictably mixed, ranging from critics suspecting an influx of imported seeds, which would compound the technological domination of the seed industry by MNCs (Bhattacharya, 1988; Menon and Sadananda, 1989), to those predicting the genesis of a resilient domestic seed industry (Pray, 1990). In contrast to Pray's predictions, imports increased dramatically: seed imports rose from 14 tonnes in 1988–89 to exceed 1600 tonnes in 1993–94, while the number of planting materials imported increased from 427 106 to exceed 11 mn in the same period (Niranjan Rao, 1997).

Whether these imports are the genesis of a period of technological dependence of the domestic seed industry is a question that is currently difficult to unambiguously answer. What is presently clear is that the seed industry has been subject to a phase of mergers and acquisitions initiated by MNCs resulting in significant consolidation: Monsanto has acquired a 26 per cent stake in Mayhco, Agrevo controls 100 percent of Proagro, Emergent Genetics has acquired a 74 per cent stake in Mahendra Hybrids and Pioneer Hybrid has a 51 per cent stake in SPIC.[6]

Recent scholarship on the private sector plant breeding activities in India can be classified into two groups: (a) macro studies of the industry based on survey data, and (b) micro studies focusing on a crop and/or region. These studies provide rich empirical observations and data that require closer analysis. Pray, Ramaswami and Kelly[7] compare and assess the performance of private investment in the Indian seed industry at two points in time, viz. 1987 and 1995, to discern the impact of the 1980s reforms. The empirical data that they present (cf. Table 7.3) provides the basis for their conclusion that the reforms have had a positive impact on private sector investments and recommend further liberalization.

Three empirical observations are central to their conclusion. First, the R&D effort (i.e. total R&D expenditures and the number of scientists employed) increased substantively across the two surveys. Since a large share, 44 per cent of total (nominal) R&D expenditures, was on the part of entrants, it is suggested that reforms stimulated private sector investments. Moreover, for incumbents, they suggest that the threat to existing seed market shares placed by the large entrants must have been a primary motivating factor; that is, an indirect consequence of the reforms.

Table 7.3 Economic characteristics of private sector seed companies

	Average sales (Rs million)	Average number of employees	Average R&D (Rs million)	R&D to sales ratio (%)	R&D to employee ratio (Rs million)
Entrants[a]					
Large (5)	174	70	10.0	5.75	0.143
Small (15)	17	14	0.7	4.12	0.050
Incumbents[b]					
Large (7)	282	313	11.0	3.90	0.035
Small (20)	41	72	0.9	2.20	0.013

Notes: The number of firms in each class are reported in brackets; the total number of firms in the survey = 47; all R&D figures are nominal. [a] Entrants are firms that entered the seed industry after 1987, totalling 20. [b] Incumbents are firms in existence prior to 1987 regulatory changes, totalling 27.

Source: Adapted from Pray, Ramaswami and Kelly (n.d.).

Second, the relatively higher R&D/sales ratio of entrants in comparison to incumbents suggests that the reforms provided incentives for greater R&D effort. Finally, the study draws attention to a fall in concentration ratios – 75 per cent of industry R&D is distributed across 10 firms, whereas in 1987 it was accounted for by three firms and four-firm concentration ratios dropped from 69 per cent (1987) to 51 per cent (1995) – suggesting that the reforms have attracted entrants and lead to a widening of the pool of technology suppliers and a broadening of the industry structure.

A closer examination of the three empirical observations is warranted. It is important to note that the study uses nominal estimates for R&D expenditures, thus raising questions on the real increase in R&D expenditure with time. Moreover, it is useful to place private sector investments in the context of total investments in agricultural research. One widely quoted estimate is that the private sector accounts for 15 per cent of total investments in agricultural research in India (Ravishankar and Archak, 2000). Second, the fact that entrants exhibit a relatively higher R&D/sales ratio in comparison to incumbents is well-established in seed industry studies of Latin America (Jaffe and Van Wijk, 1996) and the USA (Butler and Marion, 1985). This phenomenon, it has been explained, is on account of the startup costs associated with breeding programmes, the gestation lag between breeding and commercialization and the barriers to entry in the seed market. Finally, evidence of diversification of technology suppliers is promising. However, given the limited substitutability between seeds of different crops and that varieties tend to be developed for specific regions, conclusive evidence of increased competition requires data disaggregated at the level of crop and region. No doubt there has been an increase in private sector investments in plant breeding; however, the causal mechanisms are more complicated than suggested here. Further, the 'impact' of the reforms, in particular the claim that private investments have been stimulated, remains an open question.

A series of papers have reported transformations in the maize seed industry, where the following patterns are evident (Singh *et al.*, 1995; Singh and Morris, 1997; Morris *et al.*, 1998; cf. Table 7.4):

- By 1993 there were 25 maize breeding programmes in the private sector – a number comparable to the public sector. However, there are important differences in the scope and nature of breeding activities. At one level, public breeding tends to reveal higher scientific content,[8] whereas private breeding is skewed towards seed production (34%) and distribution (27%).

- The private sector has a larger share of the seed market having increased from a marginal presence in the early 1980s (less than 10%).[9] Beginning in the mid-1980s, the private sector has rapidly increased the number of varieties released, yet almost 45 per cent of private sector hybrid seed sales are of public sector hybrids.
- The private sector focuses exclusively on hybrids, while the public sector predominantly focuses on open-pollinated varieties (73%) suitable for marginal farmers.
- The private sector's dependence on the public sector extends beyond the apparent licensing of public sector hybrids and the use of germplasm from the public sector. In addition, there is an indirect *subsidy* from the public sector to the private sector in the form of human resource development.

Evidence from the rice seed industry in Andhra Pradesh is insightful because the private sector undertakes no breeding activity and the crop is not a typical high-value/low-volume crop, and yet, seed production is increasingly conducted by the private sector (Tripp and Pal, 2001). A number of factors explain this phenomenon: (a) a large captive market of 3.5mn hectare rice cultivation, with regular seed purchase by farmers,[10] (b) rice seed production is relatively simple and does not require substantial investments, (c) availability of experienced seed growers (in Northern Telangana) and storage and processing facilities, and importantly, (d) availability of breeders' seed from the public sector. Importantly, the viability of many seed companies is crucially contingent on securing

Table 7.4 The Indian maize seed industry, 1993

	Public sector	**Private sector**
Number of research programmes	27	25
Number of researchers[a]	108.5	92.0
of which[b]:		
• germplasm improvement (%)	77	92
• crop management research (%)	15	8
• other (%)	8	0
Number of varieties released (1961–93)	96	80
• of which, %age hybrids	27	100
Percentage of varieties containing		
• ICAR germplasm	100	33
• CIMMYT germplasm	35	25

Note: [a] Full-time equivalents; [b] proportion of researchers devoted to the activity.
Source: Adapted from Morris *et al.* (1998).

free access to publicly developed varieties. Clearly, the future competitiveness of the industry depends on the terms of access to the results of public sector research (Pray, 1990).

The nature of private sector entry has differed across the two crops. Private investments in hybrid maize appears to spread across from varietal development to seed processing and marketing, whereas in the rice seed industry the private sector has contained itself to seed production and distribution. Given the uneven performance of the private sector and changes in the relationship between the public and the private sector, policy questions concerning the organization of agricultural research in India need to be addressed.

The mandate for public plant breeding

The introduction of PBRs is recognized as dually generating opportunities for the public sector to reorient its funding patterns and activities while also raising concerns about the deepening institutional division of labour between the public and the private sector (Godden, 1982; Kloppenburg, 1988). Private investment in plant breeding has a propensity to focus on select crops that favour particular technological trajectories (e.g. dwarfs, chemical responsiveness, yield as opposed to quality). This is often reinforced by marketing decisions and scale economies which could promote a portfolio of marginally differentiated varieties that map across a narrow set of characteristics (Rangnekar, 2001). Consequently, some techno-economic areas (i.e. crops and production needs) will remain neglected by private investments. Hence, the policy rationale, based on the traditional logic of underinvestment by the private sector, for the public sector to withdraw from activities where the private sector exists and allocate resources in areas that remain underfunded. However, on the other hand, given the potential for adverse distributional consequences of IPRs, the rationale for public sector resource allocation is to promote access to and exchange of germplasm. These are some issues concerning the mandate for public plant breeding in an era of increasing private sector presence. Here we focus our attention on two questions:

- Setting research priorities and meeting national needs: Is it necessary for the public sector to maintain breeding programmes to supplement private sector investments and ensure that diverse national needs are met?
- Maintaining the public domain: Is it necessary to develop and maintain genetic resources and enabling technologies in the public domain so as to ensure a socially efficient use of these resources?

As evidence of the 'impact' of private sector plant breeding in India is limited, it is necessary to assess these two policy rationales in a wider context of empirical observations available from studies on PBRs and IPRs in other countries and regions.

In terms of the first question, it is a well-established fact that the focus of private sector breeding efforts is mixed and unevenly spread across select crops. This is easily borne out from empirical studies in the USA, where wheat and soybean were the main beneficiaries of private investment (e.g. Butler and Marion, 1985; Kloppenburg, 1988; Fuglie *et al.*, 1996), and Latin America, where private investment was devoted mainly to hybrids (Jaffe and van Wijk, 1996). In addition, varieties developed within private breeding programmes tend to suit exclusive agro-ecological zones where farmers have the capacity to purchase essential complementary inputs. This is evident in the maize breeding programme in India (see above). The breeding effort is not devoted to meeting the diverse and variable needs of farming communities that do not have the capacity to purchase key inputs (Hardon, 1992). Lesser (1991) agrees that the dominant agri-research paradigm for developing countries is devoted to 'post-Green Revolution' areas and a focus on select crops. Repeated experience within extension programmes have established the need for dedicated breeding strategies to develop varieties suited to the 'complex, diverse and risk-prone' setting of low-input farmers (Tripp, 1997). For example in India, commentators remain wary of the benefits of increased adoption of genetically uniform varieties, which could be disadvantageous for the biological requirements of resource-poor farmers (Wood and Bombin, 1990; Jain, 1991)

Recent assessments of the portfolio of agri-biotech research further confirms the fact that the crops and the production needs of resource-poor farmers in developing countries are neglected (Pardey *et al.*, 2001; Fresco, 2001). It is difficult not to underestimate this problem. FAO's (1998) *State of the World's Plant Genetic Resources* draws attention to the case of 'resource poor' farmers – estimated at over one-half of the global farming population[11] – who work on marginal lands and farm minor crops. These crops, which include staples like millet and groundnut and a range of fruits and vegetables, contribute substantially to their nutrition but have failed to attract the attention of crop-development programmes.

The above situation clearly warrants a public plant breeding programme. This in effect is the traditional rationale for public investment; that is, underinvestment by the private sector. Commentators emphasize the need for maintaining public breeding programmes at the international and national level to fill the dual gap left by the private sector in terms of neglected crops and production needs (e.g. Barton, 1982; Wood and Bombin, 1990; Hardin, 1992; Pardey *et al.*, 2001). An implication of this mandate for public plant breeding is that it does not effectively compete with the private sector since the markets being served are well-segmented and different. Consequently, possible disincentive effects of a public plant breeding programme on the private sector may not be expected. Yet, some public plant breeders recommend the need for maintaining healthy competition between the two (e.g. Anon, 1990). This aside, it is also suggested that public plant breeding has a national obligation to set the social research agenda (Kloppenburg, 1988). Recalling the central importance of the 'seed' and the potential for exercising control over the direction of technical change in agriculture, it is essential that breeding programmes remain within the public sector (Rangnekar, 2001).

The second question identified above relates to deeper issues concerning public-domain knowledge and the public funding of research (e.g. Stephan, 1996; Pavitt, 2000). This literature draws attention to the positive productivity influence of changes in the stock of knowledge flowing from publicly funded research. Changes in the 'stock of knowledge' generate new techno-economic opportunities that stimulate and encourage the private sector to undertake developmental research. Naturally, the direction in which basic research progresses has an important bearing on the characteristics of downstream developmental research. Studies have established this influence of publicly-funded research on private sector activities in the form of a geographical clustering of private laboratories around key university centres/departments (Audretsch and Stephan, 1996) and also through bibliometric research identifying the citation of 'star scientists' in patent documents (McMillan *et al.*, 2000). Here, we suggest that a similar dynamic exists in terms of the use of publicly-bred lines and varieties in the breeding of new cultivars by the private sector, evidence of which has been reported elsewhere in this chapter. Of similar significance, are key breeding achievements like the development of hybrid corn in the USA and dwarf varieties of the Green Revolution which occurred in the public sector.

There are other useful aspects of this role for public plant breeding. For instance, there are potentially adverse distributional consequences of the use of IPRs that arise from practices like 'patent fencing' areas of research or securing 'overlapping patents' across specific technological areas. These practices could potentially lead to the control of particular research trajectories and retard scientific progress (e.g. IDRC, 1994;

Lewontin and Santos, 1997; Heller and Eisenberg, 1998; Wright, 2000). Closely related to this are potential restrictions on access to and exchange of germplasm and related enabling technologies – the very building blocks of new cultivar development. Here consider a striking example of an open-access era of plant breeding: CIMMYT's wheat variety called VEERY (*c.* 1977) was a breeding effort involving over 3000 crosses and 51 parental lines. Problems of access arise when the scope of IPRs are uncertain and ownership is diffuse, thus making it difficult to negotiate access (Pardey *et al.*, 2001). The continued development of key resources (germplasm and enabling technologies) in the public domain will promote the viability of small seed companies and ensure a broad industrial structure. Following this exposition the need to maintain key genetic resources and related enabling technologies in the public domain can be easily appreciated. However, questions remain on how the public sector should regulate this public domain (see Godden, 1984, for a discussion).

To sum up, there are strong reasons for maintaining a public breeding programme. First, the mixed and uneven focus of private plant breeding means that certain crops and production needs will remain neglected. Second, it is necessary to develop and maintain key resources within the public domain to ensure a socially efficient use of information and diminish some of the potentially adverse implications of IPRs.

Conclusion

This chapter has examined various rationales for public plant breeding in order to present a mandate for continued public investments in plant breeding. The importance of plant breeding is explained in terms of the properties embedded in the 'seed' and the dual role of 'seeds' in delivering technical change to agriculture and interlinking markets producing inputs for agriculture. The chapter recognises the obligation under the TRIPs Agreement to introduce some form of IPRs for plant varieties. However, this still leaves open the question of what the public sector will do in the future.

Based on a review of the evidence of post-1980s reform performance of the private sector in the Indian seed industry, supplemented by evidence from other countries, two broad rationales for maintaining a public plant breeding programme are proposed. First, the limited focus of private plant breeding on select crops and specific regions means that certain crops and regions will remain neglected. This underinvestment by the private sector forms the traditional rationale for public investments in research. Second, it is contended that public investments have an important function of developing and maintaining knowledge and resources in the public domain. In the case of plant breeding it is pertinent to maintain genetic resources and associated enabling technologies in the public domain. Moreover, it is important to appreciate the wider role of public investments with respect to agriculture research, such as the training of personnel, variety testing, the collection, documentation and maintenance of genetic resources and the development of seed regulations. No doubt, the continued presence of public plant breeding, in particular varietal development, raises concerns of 'crowding out' of the private sector. To some extent these concerns are overstated as the public sector would largely focus on neglected crops and regions. However, the general concern is legitimate and requires a detailed investigation.

Notes

1 This section is based on Rangnekar (2000: chapter 6), where an expanded analytical treatment is available.
2 The term 'focusing device' is used in the sense of Rosenberg (1976) and Sahal (1990).
3 A similar argument can be made for the case of mechanization. In this sense, mechanization of farm practices requires specific transformations in the biology (e.g. uniform maturing) and architecture (e.g. hard exteriors and strong stalks) of the plant before the adoption of mechanization is economically viable.
4 Rangnekar (2001) has a detailed discussion of several policy questions identified in the IPGRI document.
5 The rhetoric of withdrawing the public sector from *near-market* activities was the cornerstone of the Thatcher government's sale of the Plant Breeding Institute, Cambridge, in 1988 (Webster, 1989).
6 I am grateful to C.S. Srinivasan for sharing these data with me, which were collected in early 2001.
7 I am grateful to Bharat Ramaswami for sharing a prepublication draft of this article with me.
8 Singh *et al.* (1995, p. 15) note that public breeding in India does not have a high 'scientific component' relative to global averages, but the results demonstrate a 'relatively sophisticated inbreeding and test-crossing, attesting to their considerable resource endowments and high levels of human capital'.
9 An important factor contributing to the increasing penetration of the private sector in the seed market is the behaviour of farmers, in particular their decisions concerning adoption of hybrids and seed saving. In areas where corn is a commercial crop, e.g. Andhra Pradesh, Karnataka and Bihar, farmers have tended to be quick adopters of hybrids and reveal high seed replacement rates (Singh and Morris, 1997).
10 Tripp and Pal (2001) indicate that between 20–40% of seed is farm-saved seed.
11 A very large of these resource poor farmers are in India.

References

Anon. (1990) *Report of the Sub Group V of the Standing Policy Planning Committee of the Governing Body of ICAR on Gene Patenting and Breeders' Rights.* New Delhi: Indian Council of Agricultural Research.

Audretsch, D.B. and Stephan, P.E. (1996) 'Company–Scientist Locational Links: The Case of Biotechnology', *American Economic Review*, vol. 86(3), pp. 641–52.

Barton, J.H. (1982) 'The International Breeder's Rights System and Crop Plant Innovation', *Science*, vol. 216(4), pp. 1071–5.

Bhattacharya, A. (1988) 'New Seed Policy: Whose Interest Would it Serve', *Economic and Political Weekly.*

Bingham, J. and Austin, R.B. (1993) 'Achievements and Limitations of Wheat Breeding in the United Kingdom', in D.R. Buxton, R. Shibles, B.L. Forsberg, B.L. Blad, K.H. Asay, G.M. Paulsen and R.F. Wilson (eds), *International Crop Science Congress I.* Madison, WI: Crop Science Society of America, pp. 181–7.

Brockway, L. (1979) *Science and Colonial Expansion: The Role of the British Royal Botanical Gardens.* London: Academic Press.

Butler, L.J. and Marion, B.W. (1985) *The Impacts of Patent Protection on the US Seed Industry and Public Plant Breeding* (North Central Regional Research Publication no. 304). Madison: Research Division, College of Agricultural and Life Sciences, University of Wisconsin-Madison.

Bye, P. and Fonte, M. (1994) 'Is the Technical Model of Agriculture Changing Radically?' in P. McMichael (ed.), *The Global Restructuring of Agro-food Systems.* Ithaca: Comell University Press, pp. 241–57.

Cohen, J.I. (1999) 'Managing Intellectual Property – Challenges and Responses for Agricultural Research Institutes', in J.I. Cohen (ed.), *Agricultural Biotechnology and the Poor: Addressing Research Program Needs and Policy Implications.* Wallingford, Oxon: CAB International, pp. 209–17.

Cromwell, E. (1990) *Seed Diffusion Mechanisms in Small Farmer Communities: Lessons from Asia, Africa and Latin America* (vol. ODI Agricultural Administration Network Paper 21). London: Overseas Development Institute.

Dhar, B. and Chaturvedi, S. (1998) 'Introducing Plant Breeders' Rights in India: A Critical Evaluation of the Proposed Legislation', *Journal of World Intellectual Property*, vol. 1(2), pp. 245–62.

Dhar, B., Pandey, B. and Chaturvedi, S. (1995) 'Farmers' Interests Recognized in Indian PBR Bill', *Biotechnology and Development Monitor*, vol. (23), pp. 18–21.

Evenson, R.E. and Putnam, J.H. (1987) 'Institutional Change in Intellectual Property Rights', *American Journal of Agricultural Economics*, vol. 69(2), pp. 403–9.

FAO (1998) *The State of the World's Plant Genetic Resources for Food and Agriculture.* Rome: FAO.

Fowler, C. (1994) *Unnatural Selection: Technology, Politics and Plant Evolution.* Switzerland: Gordon and Breach Science Publishers SA.

Fresco, L.O. (2001) *'Genetically Modified Organisms in Food and Agriculture: Where are We'? Where are we going?,* Transcript of a talk delivered at the FAO.

Fuglie, K., Ballenger, N., Day, K., Klotz, C., Ollinger, M., Reilly, J., Vasavada, U. and Yee, J. (1996) *Agricultural Research and Development: Public and Private Investments Under Alternative Markets and Institutions* (Agricultural Economics Report 735). Washington, DC: Economic Research Service, Department of Agriculture.

Godden, D. (1982) 'Plant Variety Rights in Australia: Some Economic Issues', *Review of Marketing and Agricultural Economics*, vol. 50(1), pp. 51–95.

Godden, D. (1984) 'Plant Breeders' Rights and International Agricultural Research', *Food Policy*, pp. 206–18.

Goodman, D. and Redclift, M. (1991) *Refashioning Nature: Food, Ecology and Nature*. London: Routledge.

Goodman, D., Sorj, B. and Wilkinson, J. (1987) *From Farming to Biotechnology*. Oxford: Blackwell Publishers.

Gregg, B.R., Delouche, J.C. and Bunch, H.D. (1980) 'Inter-relationships of the Essential Activities of a Stable, Efficient Seed Industry', *Seed Science and Technology*, vol. 8, pp. 202–27.

Hardon, J.J. (1992) 'Biotechnology, Plant Breeding and Resource Poor Farmers in the Third World', In H. Brouwer, E.M. Stokhof and J.F.G. Bunders (eds), *Biotechnology and Farmer's Rights: Opportunities and Threats for Small Scale Farmers in Developing Countries*. Amsterdam: VU University, pp. 73–7.

Heller, M.A. and Eisenberg, R.S. (1998) 'Can Patents Deter Innovation? The Anticommons in Biomedical Research', *Science*, vol. 289, pp. 698–701.

IDRC (International Development Research Centre). (1994) 'Plant Breeders Rights (A Position Paper by IDRC)', Reprinted in Anon. (ed.), *Breeders' and Farmers' Rights: An Interdisciplinary Dialogue. Background Papers*. Madras: M.S. Swaminathan Research Foundation (Centre for Research on Sustainable Agricultural and Rural Development), pp. 93–114.

Jaffe, W. and Van Wijk, J. (1996) *The Impact of Plant Breeders' Rights in Developing Countries: Debate and Experience in Argentina, Chile, Colombia, Mexico and Uruguay*. The Hague: Ministry of Foreign Affairs.

Jaffee, S. and Srivastava, J. (1992) *Seed System Development: The Appropriate Roles of the Private and Public Sector*. Discussion Paper no. 167. Washington, DC: World Bank.

Jaffee, S. and Srivastava, J. (1994) 'The Roles of the Private and Public Sector in Enhancing the Performance of Seed Systems', *The World Bank Research Observer*, vol. 9(1), pp. 97–117.

Jain, H.K. (1991) 'Plant Genetic Resources and the Efficiency Factor in Agriculture', In I.P. Getubig, jr. and M.S. Swaminathan (eds), *Biotechnology for Asian Agriculture: Public Policy Implications*. Kuala Lampur: Asian and Pacific Development Centre, pp. 95–104

James, C. (1998) 'Global Status and Distribution of Commercial Transgenic Crops in 1997', *Biotechnology and Development Monitor*, vol. 35, pp. 9–12.

Kloppenburg, J. Jr. (1988) *First the Seed: The Political Economy of Plant Biotechnology*. Cambridge: Cambridge University Press.

Knudson, M. (1990) 'The Role of the Public Sector in Applied Breeding R&D', *Food Policy*, vol. 15(June), pp. 209–17.

Lesser, W.H. (1991) *Equitable Patent Protection in the Developing World: Issues and Approaches*. Tskuba: Eubios Ethics Institute.

Lewontin, R.C. and Berlan, J.-P. (1990) 'The Political Economy of Agricultural Research: The Case of Hybrid Corn', In C.R. Carroll, J.H. Vandermeer and P.M. Rosset (eds), *Agroecology*. New York: McGraw Hill, pp. 613–28.

Lewontin, R.C. and Santos, M.d.M. (1997) 'Current Trends in Intellectual Property Rights Protection Pose Serious Threats to Future Innovations in Agricultural Sector', *Diversity*, vol. 13(2 & 3), pp. 25–7.

McMillan, G.S., Narin, F. and Deeds, D.L. (2000) 'An Analysis of the Critical Role of Public Science in Innovation: the Case of Biotechnology', *Research Policy*, vol. 29(1), pp. 1–8.

Menon, U. and Sadananda, A.R. (1989) *The Seeds of Dependence: A Close Look at the New Seed Import Policy*. New Delhi: Delhi Science Forum.

Mooney, P.R. (1983) 'The Law of the Seed: Another Development and Plant Genetic Resources', *Development Dialogue* vol. (1–2), pp. 1–173.

Morris, M.L., Singh, R.P. and Pal, S. (1998) 'India's Maize Seed Industry in Transition: Changing Roles for the Public and Private Sectors', *Food Policy*, vol. 23(1), pp. 55–71.

Mruthyunjaya and Ranjitha, P. (1998) 'The Indian Agricultural Research System: Structure, Current Policy Issues, and Future Orientation', *World Development*, vol. 26(6), pp. 1089–101.

Niranjan Rao, C. (1997) *Plant Variety Protection and Plant Biotechnology Patents: Options for India*. New Delhi: Allied Publishers Ltd in collaboration with UNDP.

Pardey, P.G., Wright, B.D. and Nottenburg, C. (2001) 'Are Intellectual Property Rights Stifling Agricultural Biotechnology in Developing Countries?' In Anon. (ed.), *IFPRI 2000–2001: Annual Report*. Washington, DC: IFPRI.

Perrin, R.K. (1994) 'Intellectual Property Rights in Agricultural Development', In J.R. Anderson (ed.), *Agricultural Technology: Policy Issues for the International Community*. Washington, DC: Agricultural Policies Division, The World Bank.

Pray, C., Ramaswami, B. and Kelley, T. (n.d.) 'The Impact of Economic Reforms on R&D by the Indian Seed Industry', *Food Policy*.

Pray, C.E. (1990) 'The Potential Impact of Liberalising India's Seed Laws', *Food Policy*, vol. 15(3), pp. 193–98.

Pray, C.E. and Ramaswami, B. (1991) *A framework for seed policy analysis in developing countries*. Washington, DC: International Food Policy Research Institute.

Ramaswami, B. (2000) *Intellectual Property Rights in Agriculture: Implications and Challenges for Public Research and Policy*, Indian Statistical Institute, mimeo.

Rangnekar, D. (1998) 'Tripping in Front of UPOV: Plant Variety Protection in India', *Social Action*, vol. 48(4), pp. 432–51.

Rangnekar, D. (2000) *Innovative Appropriation: The Role of Economics, Science and Law in Transforming Plants into Products, A Case Study of Wheat Breeding in the UK*. Unpublished Dissertation, School of Economics, Kingston University, Kingston, UK.

Rangnekar, D. (2001) 'Access to Genetic Resources, Gene-Based Inventions and Agriculture'. London: Study paper commissioned by the UK government's Commission on Intellectual Property Rights. Available at www.iprcommission.org

Ravishankar, A. and Archak, S. (2000) 'Intellectual Property Rights and Agricultural Technology-Interplay and Implications for India', *Economic and Political Weekly*, vol. 35(27), pp. 2446–52.

Rosenberg, N. (1976) 'On Technological Expectations', *Economic Journal*, vol. 86(343), pp. 523–35.

Sahal, D. (1990) 'Technological Guideposts and Innovation Avenues', In C. Freeman, (ed.), *The economics of innovation. International Library of Critical Writings in Economics, no. 2*, Aldershot, UK and Brookfield, Vt: Elgar, pp. 442–63.

Singh, R.P., Pal, S. and Morris, M.L. (1995) *Maize research, development, and seed production in India: Contributions of the public and private sectors.* Economics Working Paper no. 95–03. Mexico: CIMMYT.

Singh, R.P. and Morris, M.L. (1997) *Adoption, management, and impact of hybrid maize seed in India.* Economics Working Paper no. 97–06. Mexico: CIMMYT.

Stephan, P.E. (1996) 'The Economics of Science', *Journal of Economic Literature*, vol. 34(3), pp. 1199–235.

Tripp, R. (1997) 'Introduction', In R. Tripp (ed.), *New Seed and Old Laws: Regulatory Reform and the Diversification of National Seed Systems.* London: Intermediate Technology Publications on behalf of the Overseas Development Institute, pp. 3–13.

Tripp, R. and Pal, S. (2001) 'The Private Delivery of Public Crop Varieties: Rice in Andhra Pradesh', *World Development*, vol. 29(1), pp. 103–17.

Webster, A. (1989) 'Privatisation of Public Sector Research: The Case of a Plant Breeding Institute', *Science and Public Policy*, vol. 16(4), pp. 224–32.

Wood, D. and Bombin, L.M. (1990) *Plant breeders' rights in India: A (draft) report to the Government of India* (TCP/IND/0052(A)). Rome: Food and Agriculture Organisation of the UN.

Wright, B.D. (2000) 'International Crop Breeding in a World of Proprietary Technology', In V. Santaniello, R.E. Evenson, D. Zilberman and G.A. Carlson (eds), *Agriculture and Intellectual Property Rights: Economic, Institutional and Implementation Issues in Biotechnology.* Wallingford, OXON: CAB International, pp. 127–38.

8

The Rise of Entrepreneurship in India

Dipendra Sinha

Introduction

While economists have long recognized the role of entrepreneurs in economic growth, Schumpeter was the first economist to formalize the myriad roles that they play. Entrepreneurs have the following functions according to Schumpeter (1934 and 1939). First, they have the task of reorganizing industry; second, they exploit new sources of inputs; third, they introduce new technologies to reduce the inputs needed to produce a given amount of output; fourth, they open up markets; and fifth, they introduce new products.

In his *Principles of Economics* (1890), Alfred Marshall distinguished between four factors of production, namely labour, capital, land and entrepreneur. Entrepreneurship remains an elusive concept. As Leff (1979) pointed out, the term has even been used to denote a 'firm' or management in general. What are the special qualities of entrepreneurs? According to Frank Knight (1921), they have special abilities for taking risks and dealing with uncertainties; the entrepreneur is the ultimate decision-maker in the enterprise. Hired managers are not entrepreneurs because they do not commit ownership and do not bear risks. In developing countries, entrepreneurs may face even more challenges, since markets for bearing risk and uncertainty are even less complete. Due to rapid structural changes and poor information, the absolute amount of uncertainty may be greater in developing countries (Leff, 1979). Also, in developing countries entrepreneurs may not get the type of support they would expect to have in countries where market capitalism is more highly developed, and where credit markets are more responsive to the entrepreneur's needs. Thus, in many developing countries, special measures are taken to encourage entrepreneurial activities.

Evidence gathered by McClelland (1961) and McClelland and Winter (1971) suggests that the need for achievement is one of the crucial factors in the development of entrepreneurs. This suggests that how children are reared has an effect, which has relevance for developing countries. Even in non-communist developing countries, traits of individualism are often not encouraged, but individualism is often conducive to independent decision-making and thus to entrepreneurship. McClelland believes that elaborate training for entrepreneurs can help them to succeed.

As Kilby (1983) points out, earlier studies on entrepreneurship involved interviews with entrepreneurs. Performance indicators such as size of assets, sales, employment and profits were related to education, occupational background, social status, political influence, religious affiliation, ethnic origin and different economic variables. Studies show that ethnic and religious factors are not important determinants of performance, which implies that the entrepreneurs could be drawn from various segments of society. One curious finding of many of these studies is that education is also not an important determinant. Economic variables, on the other hand, are found to be important. Kilby notes a number of limitations of such research. First, the data on performance indicators may not be very reliable. Second, in many cases companies may depend a lot on sales to the government sector. If the government does not place orders, some companies may not even survive. Third, the data do not include failed entrepreneurs, and, also the upper reaches of the manufacturing sector are not included in these studies. In the latter, such as multinational corporations, the state, joint enterprises and alien minorities tend to dominate. The case of Chinese Malaysians in Malaysia is a case in point.

Leff (1979) takes a different track. He argues that several things point to entrepreneurship becoming less of a bottleneck in developing countries. First, many developing countries have been growing at a fast rate, which shows that an inelastic supply of entrepreneurs may not be the case in those countries. Second, developing countries have been more careful in screening foreign investment proposals, and, for example, have limited the participation of multinational corporations to high-technology activities. Restrictions placed on alien minorities and expulsions and at times, massacres of these minorities in African and Southeast Asian countries imply that there is no dearth of indigenous entrepreneurs. Third, there has been a decrease in the number of journal articles and books on the subject of entrepreneurship; recent editions on development economics have devoted less attention to entrepreneurship than earlier editions.

Leff's arguments can be criticized on a number of grounds. First, one can argue that some countries have opened up even more than before to foreign investment. Since the opening up in 1991, India is now more open to foreign investment than anytime since independence. Also, for India, the alien minority is much less important. Parsis (those who migrated from Iran centuries ago) have provided many entrepreneurs, but unlike alien minorities in many other countries they have assimilated well into the mainstream of Indian society, except in the case of marriage – Parsis tend to marry within their own communities. Parsis have not been the subjects of persecution in India. Second, the growth rate in a country is a function of a number of factors – not just entrepreneurs.

What determines the choice that a person makes whether to be an entrepreneur or a worker? According to Khilstrom and Laffont (1979), more risk-averse people choose to be workers, while those less risk-averse choose to be entrepreneurs. However, according to Banerjee and Newman (1993), initial capital is the determining factor.

History of entrepreneurship in India

Indians have never lacked the spirit of entrepreneurship, and from time immemorial Indian products have made their way to other countries. Indian muslin was used by the Pharaohs of Egypt. The managing agency houses wielded a lot of power and prestige during the first half of the twentieth century (Lokanathan, 1935), and were similar to holding companies. They had control over the three most important industries in eastern India – jute mills, collieries and tea plantations. In contrast, in the western, northern and southern parts of India, Indian entrepreneurs dominated. For example, in western India most entrepreneurs were Parsis, Gujaratis, Cutchi Memons and Sindhis (Goswami, 1989). Also, certain groups have been more active entrepreneurs than other groups. Marwaris from the Marwar region of the state of Rajasthan formed a well-known business class especially in Kolkata (formerly, Calcutta) which used to be the capital of India until 1911. In western India, Jains (a religious sect) have been in the forefront of businesses.

British rule in India was marred by discrimination against Indian entrepreneurs. For example, Bengali entrepreneurs were discriminated against in the 1800s because they competed with English entrepreneurs in international commerce. Such discrimination often led to the withdrawal of reputed local firms from the business sector (Dana, 2000). Thus, British rule was not conducive to the development of entrepreneurship in India. Nafziger (1971) attributes the lack of indigenous entrepreneurs partly to the British policy of discrimination and duplicity.

India's manufacturing sector accounted for only 17 per cent of India's national income and employed about 10 per cent of the workforce in the country in 1950 (Goswami, 1989). Even by 1991, the manufacturing sector including mining and electricity accounted for about 25 per cent of India's national income (Central Statistical Organization, 1997). When India gained independence, India started to follow a policy of a 'socialistic pattern of society'. The first Prime Minister, Jawaharlal Nehru, was very impressed with the Soviet model of planning and was eager for India to follow such a path, and the government sector wanted to occupy the 'commanding heights of the economy'. However, things did not change overnight. Even by 1957, the private share of total paid-up capital of companies at work was about 94 per cent. However, the share steadily declined until 1987, and by 1996 was around 50 per cent (Confederation of Indian Industry, various years). The restrictions on the private sector steadily increased. These restrictions not only thwarted the development of entrepreneurship in India, but also created opportunities for rent-seeking by government officials. Some changes were brought about in the 1980s, but the pace of change was rather slow.

Bottlenecks to the development of entrepreneurship in India

First, the belief system in India regards the business profession as a dirty profession. Most people prefer regular jobs with guaranteed salaries than the uncertainties associated with owning businesses. However, this attitude is not restricted to India. Especially prior to 1991, most businesspersons had to grease the palms of government servants. Thus, the belief was that owning businesses would involve indulging in corrupt practices. The British period was characterized by discrimination against Indians in businesses, especially for some ethnic groups, and in addition the maze of licensing and the control of many activities by the government was not conducive for risk-takers. Since 1991, however, things have been changing slowly. Many of the previous restrictions of the government are a thing of the past.

Second, many Indians are still fatalistic. The holy book of Gita teaches that people have rights over their work but not over the fruits of their work. The creation of wealth is not accorded a high priority by most people, and for many the belief is that it is not necessary to make an excessive amount of money. People tend to be apologetic about making money. Even the wealthy try not to show off in some cases (of course, the other reason for not showing off is to keep the tax people at bay – many businesspersons try to hide their true income and thus reduce their tax burden). Attitudes, however, do change. Recent years have brought changes in attitudes especially among the younger generation, many of whom are not that apologetic about making money. As the Indian economy opens up even more, and with increased globalization, it is expected that people's attitudes will change even more. The media, especially television, plays a significant role. Not long ago, Indians could watch only programmes on government-owned channels; now, Indians have the choice of watching a variety of channels which show programmes made in India and elsewhere. Soap operas, made in India and other countries, definitely change attitudes. Narayana Murthy, the Chairman and Managing Director of Infosys Technologies Ltd, is of the view that it is only recently that there has been some acceptance of the idea that the creation of wealth is the only way to solve the problem of poverty in India. He further suggests that the government's job is not to create wealth, but to create an environment where wealth can be created.

Third, many traditional businesses in India are run by the extended family; generations may be involved in the same business. Entrepreneurial parents are careful in choosing the type of education for their children which will be conducive to the development of faculties important for entrepreneurship. Each son and, increasingly, daughters may receive different types of education (all relevant to business) so that they can specialize in different areas in the family business. In addition, business families may often arrange marriages with other business families to further their mutual prospects (Nafziger, 1975). However, there is a downside to family entrepreneurs. They may be more conservative in taking risks, in being more innovative and in delegating authority. They may also be more reluctant to hire professional managers, although more family enterprises are now doing so.

According to a recent joint study by the Confederation of Indian Industry (CII) and German-based management consultant Ronald Berger, entrepreneurship in India is higher than in Europe (*Businessline*, January 1998). However, innovations which, according to the study, involve successful application of new ideas are lacking. Innovations need to be spurred by open competition. The study also looks at the average age of products in different countries. In Europe, the products of nearly 80 per cent of companies have an average age of less than five years. In India, on the other hand, less than 10 per cent of products have an average age of less than five years. According to Professor S.K. Bhattacharya of Warwick University, liberalization is definitely aiding innovations, but remaining bureaucracy is hurting innovations. Bhattacharya also feels that protectionist policies in international trade do not help innovations.

Caste and entrepreneurship

A number of studies have looked at the relationship between caste and entrepreneurship (see Sinha, 1995, for details). The study by Singh (1985) is based on interviews of 100 entrepreneurs of a small town in Uttar Pradesh known for its manufacturing and export of wool carpets. The study found that religion and caste had little influence on the choice of occupations. The joint family was not found to be inhibiting the growth of entrepreneurship. Also, there was a clear-cut tendency to shift away from traditional occupational obligations.

Assuming that a sound methodology was used and the interviewees revealed the truth, the study casts doubts on the many widely held beliefs about the effects of religion, caste and the joint family. Even though the study was confined to a small town in Uttar Pradesh, the findings may be true of many other small towns at least in the northern part of India.

Other studies have found that entrepreneurs tend to come from lower castes. *Aggarwal Banias* belonging to the Vaishya caste, are well-known to have produced many entrepreneurs. However, in a study of south-Indian industrialists in Vishakhapatnam, Nafziger (1975) found that upper castes were more than proportionately represented among entrepreneurs. On the other hand, Berna (1960), in a study of manufacturing enterprises in light manufacturing industry in Madras and Coimbatore found that traditional occupations and caste played a very small role in determining entry into entrepreneurial endeavours.

Economic reforms and Chambers of Commerce

According to some political commentators such as Sridharan (1993), the economic liberalization in India which started in 1991 (see Sinha and Sinha, 1994 for details of the liberalization process) was previously thought unlikely to be initiated or implemented. Previous attempts at reforms had failed, and observers had often blamed the Indian business community for such failures. It has been argued that Indian businesses which grew up under protection were not really enthusiastic about reforms which would change the character of Indian industry by bringing in competition.

Pedersen (2000) discusses various theories as to what led to the liberalization on a fairly large scale in India in 1991. The first theory is that international organizations, especially the IMF and the World Bank, were instrumental in bringing about the changes through the use of 'conditionality' in promoting policies for structural adjustments. This view of external compulsion was also propagated by the Government of India (1992). One reason for doing so might have been, in fact, to provide a justification of the reforms to those who were fully opposed to such liberalization.

The second theory (to which Pedersen seems to subscribe) is that it is the new entrepreneurs who were behind the changes. Even though the family-based industrial conglomerates (such as the House of Birlas) controlled a substantial portion of Indian industry, a technologically more advanced segment consisting of all sizes – small, medium and large – has emerged and grown more rapidly than the traditional small-scale industries. The share of the small and medium-sized technologically more advanced industries in production and export has been growing. The modern sector of the Indian economy consists not just of new companies – even the established companies have undergone changes.

These changes were reflected in the ascent of the Confederation of Indian Industry (CII) as one of the premier organizations of Indian industry. Traditionally, India had two organizations which both started before independence – the Federation of Indian Chambers of Industry and Commerce (FICCI), and the Associated Chambers of Commerce and Industry (Assocham). FICCI, dominated by the Eastern group of companies (especially the House of Birlas) generally consisted of Indian entrepreneurs while Assocham represented the British and other foreign-controlled enterprises. Since FICCI was dominated by businesses that originated in India, some Mumbai-based business groups joined Assocham. There were allegations of favouritism within FICCI, and eventually in 1974 the Association of Indian Engineering Industry (AIEI) was formed. Engineering companies, domestic and foreign, joined AIEI, and public and private sector companies belonging to both small and large-scale industries also joined. AIEI was renamed the Confederation of Engineering Industry (CEI) in 1986, and finally as the Confederation of Indian Industry (CII) in 1992.

As pointed out by Pedersen (2000), CII is a new breed of organizations in India. Its membership includes organizations which are small, medium and large in terms of employment, and its reported philosophy is deregulation and delicensing in all areas. It also supports globalization.

As Kochanek (1996) notes, CII is in favour of working with the government in achieving its goals. CII's Secretary General, Tarun Das, has been in the forefront of public discussions on economic reforms. Pedersen feels that the rise of CII signifies not only the emergence of a new breed of entrepreneurs, but also the changes in the convictions and interests of private industry. The rise of CII, however, has not led to a massive decline of the FICCI. The FICCI was instrumental in changing government policies during the early 1990s which were perceived to be systematically favouring foreign companies in the rules governing the acquisition of preferential shares in Indian companies.

Nafziger and Rao (1996–97) studied the effects of economic reforms in the 1990s on industrial entrepreneurs in Vishakhapatnam, showing that the effects of the reforms was mixed. Generally speaking, the regulated regime suited entrepreneurs who could derive economic rents from licensing and other regulations, whilst the new regime favours entrepreneurs who are more innovative.

Entrepreneurship, saving and investment rates in India

Entrepreneurship depends crucially upon the availability of financial resources of a country. Figure 8.1 shows the total gross saving rates in India, Pakistan and Singapore during 1960–98. Total gross saving is

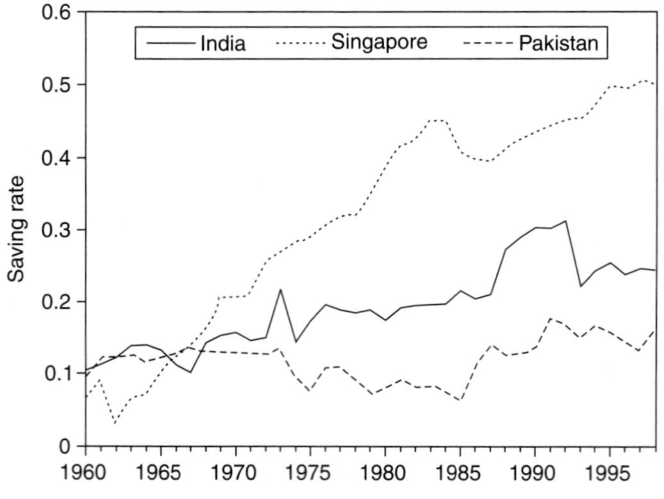

Figure 8.1 Total gross saving rates in India, Pakistan and Singapore, 1960–98
Source: International Monetary Fund (2000).

defined as GDP minus consumption (private and public consumption), and the total gross saving rate is simply gross saving divided by GDP. The figure shows the remarkable rise in Singapore's saving rate which used to be lower than that of both India and Pakistan until 1965. In 1998, it was about 50 per cent. The figure also shows that India's saving rate, which used to be about the same as that of Pakistan until about 1965, has far surpassed Pakistan's saving rate. For India, the saving rate reached the all-time high of more than 31 per cent in 1992; in 1998 it stood at about 24 per cent. India's saving rate in recent years has far exceeded the rates of many other developing countries.

Figure 8.2 shows the total gross saving rate and the gross investment rate (gross fixed capital formation as a percentage of GDP) for India during 1960–98. It is not sufficient to have a high saving rate; it must translate into a high investment rate. Except for the late 1980s and early 1990s, the saving and investment rates in India tracked each other closely since 1974. Feldstein and Horioka (1980) would argue that a high correlation between the two rates would indicate capital immobility. However, other studies since then have shown that it is possible to have a high degree of correlation between the two rates on the one hand, and a high degree of capital mobility on the other.

The *Global Competitiveness Report 2000* provides rankings of 59 countries. The two main indices are the current competitiveness index and the growth competitiveness index. For the overall growth competitiveness index, India's position climbed from 52 in 1999 to 49 in 2000. Among

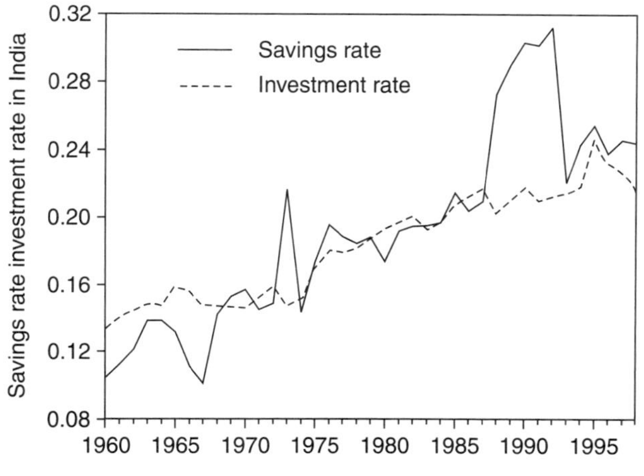

Figure 8.2 Saving and investment rates in India, 1950–98
Source: International Monetary Fund (2000).

Table 8.1 Industrial investment intentions filed through industrial entrepreneurs' memorandum and letters of intent

	Industrial entrepreneurs' memorandum			Letters of intent		
	Number	Proposed investment (Rps billion)	Proposed employment (000s)	Number	Proposed investment (Rps billion)	Proposed employment (000s)
1991	3084	763.10	769	195	20.71	34
1992	4860	1158.72	923	620	139.94	97
1993	4456	639.76	703	528	128.45	100
1994	4664	887.71	829	546	179.37	130
1995	6502	1255.09	1114	355	142.65	91
1996	4825	732.78	696	522	299.32	181
1997	3873	523.79	522	321	95.28	96
1998	2889	573.85	521	145	32.74	27
1999	2948	1288.92	477	132	8.27	17

Source: Government of India, *Economic Survey 2000–2001*. New Delhi: Ministry of Finance, 2001.

the growth competitive advantages, India ranked 14th with respect to its investment rate and 20th with respect to the national saving rate.

Table 8.1 shows the industrial investment intentions filed through the Industrial Entrepreneurs' Memorandum (IEM) and Letters of Intent (LOI). The number of IEMs increased from 3084 in 1991 to 6502 in 1995, but decreased to 2948 in 1999. The volume of proposed investment increased from Rps 763.10 billion to Rps 1288.92 billion. The proposed investment under LOIs increased from Rps 20.71 billion in 1991 to Rps 299.32 billion in 1996. However, it fell to Rps 8.27 billion in 1999.

Small scale industry policy in India

Has the Small-Scale Industry (SSI) policy worked in India? This policy of 1991 visualized liberalization of the industrial sector. A study conducted by the Institute of Small Enterprise Development (ISED) of Kerala found that, in Kerala, the industrial structure reveals a case of 'dependent entrepreneurship' (*Businessline*, 17 November 2000). In other words, entrepreneurs are becoming more and more dependent upon the state government, political parties and promotional agencies. These organizations increasingly determine the actions of entrepreneurs. While the policy of *Panchayati Raj* advocated policies of decentralization, in practice the institutions of *Panchayati Raj* have added to the dependency of entrepreneurs. According to the Director of ISED, P.M. Mathew, Kerala's industrial structure is characterized by (a) dominance of service industries, (b) a heavy bias of agro-processing activities in the manufacturing sector, and (c) a significant growth of micro enterprises and activities. In Kerala, 78.2 per cent of all units are accounted for by four categories of industries: food products, general engineering, printing and publishing, and automobiles and other services. The service industries are characterized by a low level of sustainability, and the base of service industries is likely to be substantially eroded as India's middle class grows, and as, with increased globalization, people's tastes and preferences change more towards the products of multinational corporations.

The growth of women entrepreneurs

The issue of women entrepreneurs has received worldwide attention in recent years. Ten-year data for the United States show that between 1987 and 1996, women started businesses at twice the rate of men; women-owned businesses grew by 78 per cent during the period. The 1991 Census in India showed that women entrepreneurs accounted for 10 per cent of all businesses, and estimates for 2000 puts this figure at 20 per cent. The service sector, beauty and skin-care outfits and retailing constitute the majority of these businesses. In the state of Karnataka, the Association of Women Entrepreneurs of Karnataka (AWAKE) is helping potential women entrepreneurs through its 'business incubator' programme. This programme tries to make entrepreneurs out of women by training them and exposing them to the business world before they start their own businesses. The programme is solely confined to the food-processing sector. AWAKE members are involved with handicrafts, hotels, industrial painting, interior decorations, pharmaceuticals, printing and publicity, leather and chemicals.

The future of Indian entrepreneurship

The future for Indian entrepreneurship looks more promising than ever before. The liberalization that started in 1991 now seems irreversible, and the mindset of a substantial percentage of Indians has changed. During the decades of the 1960s, 1970s and the 1980s, many used to look to the government for solutions, and there was a sense of helplessness and frustration. This has given way to hope, and infact Indian entrepreneurs are now being sought by other countries. For example, a delegation from Botswana visited Mumbai in May 1999 to encourage Indian entrepreneurs to invest in textiles, the assembly of cars, jewelry, engineering products, printing and publishing, ceramics, glass products, electronic goods and pharmaceuticals. Botswana imposes no tariffs or quotas for the manufacturing sector, and corporate tax at a uniform rate of only 15 per cent is the lowest in southern Africa.

The Namibian President, Dr Sam Nujoma has invited Indian entrepreneurs to invest in such diverse sectors as textiles, clothing apparel, garments, leatherwear, domestic utensils, plastic moulding and pharmaceuticals, agronomic and agro-industrial sectors, the processing of precious and semi-precious stones and the hospitality industry (lodges, hotels and Indian restaurants). Namibia's trade alliance within the Southern Africa Customs Union and Common Market for Eastern and Southern Africa provides access to a huge market. A joint venture project of the Namibia Development Corporation and a major company in the Punjab to develop one of the rich agricultural regions in Namibia has already materialized.

Likewise, the Sharjah Free Zone Authority of the government of Sharjah has invited Indian entrepreneurs to invest in areas such as petrochemicals and agribusiness. Sharjah has no corporate tax, and in addition there are incentives such as 100 per cent ownership, exemptions from import and export tax and all commercial levies, and repatriation of capital and profits. Leases are for 15 years and are renewable for another 15 years, and commercial, industrial or service licenses are guaranteed in 24 hours.

The Government of India is taking a number of measures to spur the growth of the industrial sector which will help the future development of entrepreneurship in India. Recently, an Expert Group was constituted to review laws and regulations and procedures (Government of India, 2001). The Group has recommended that a new Industry Act should be enacted to focus on promotion and development of industry instead of regulation. The government is considering the feasibility of enacting the new Act, and the budget for 2000–01 announced a number of industrial policy initiatives which included the following. First, excise duty has been rationalized by the introduction of central value added tax (CENVAT) and the reduction of the number of rates of excise duty. Second, the maximum retail price (MRP)-based excise duty has been extended to an increasing number of items. This will lead to simplification of the excise procedure. Third, venture capital funds (VCF) now have a single regulator – the Security and Exchange Board of India (SEBI). Fourth, the limit for providing collateral to obtain finance has been raised from Rps 100 000 to Rps 500 000 for small-scale industry (SSI) units. Fifth, the import duty on cinematic camera and other related equipment for use in the entertainment industry has been reduced. Sixth, custom duty on several items of the IT and telecommunications sector has been reduced.

Conclusion

The importance of entrepreneurship in the economic development of a country can hardly be overemphasized. In this chapter we have looked at a number of issues pertaining to the history, development, bottlenecks and government policies with regard to entrepreneurship. We have limited our discussion to industrial entrepreneurs. The economic reforms in India since 1991 have certainly been conducive to the development of entrepreneurship.

References

Banerjee, A.V. and Newman, A.F. (1993) 'Occupational Choice and the Process of Development', *Journal of Political Economy*, vol. 101 (April), pp. 274–98.

Berna, J.J. (1960) *Industrial Entrepreneurship in Madras State*. Bombay: Asia Publishing House.

Businessline (1998) 'India High on Enterprise, Low on Innovation', 19, January.

Businessline (2000) 'India: Kerala's Industrial Scene, A Clear Case of Dependent Entrepreneurship', 17, November.

Central Statistical Organisation (1997) *National Account Statistics*. New Delhi: Central Statistical Organisation.

Confederation of Indian Industry (various years) *Handbook of Statistics*. New Delhi: CII.

Dana, L.P. (2000) 'Creating Entrepreneurs in India', *Journal of Small Business Management*, vol. 38 (April), pp. 86–91.

Feldstein, M. and Horioka, C. (1980) 'Domestic Saving and International Capital Flows', *Economic Journal*, vol. 90 (June), pp. 314–29.

Porter, M., Warner, A. and Sachs, J. (eds) (2000) *Global Development Report 2000*, New York and London: McGraw Hill.

Goswami, O. (1989) 'Sahibs, Babus and Banias: Changes in Industrial Control in Eastern India, 1918–50', *The Journal of Asian Studies*, vol. 48 (May), pp. 289–309.

Government of India (1992) *Economic Survey 1991–92*. New Delhi: Ministry of Finance.

Government of India (2001) *Economic Survey 2000–2001*. New Delhi: Ministry of Finance.

International Monetary Fund (2000) *International Financial Statistics*. CD-ROM version, December.

Khilstrom, R.E. and Laffont, J.J. (1979) 'A General Equilibrium Entrepreneurial Theory of Firm Formation Based on Risk Aversion', *Journal of Political Economy* vol. 87 (August), pp. 719–48.

Kilby, P. (1983) 'The Role of Alien Entrepreneurs in Economic Development: An Entrepreneurial Problem', *American Economic Review*, vol. 73 (May), pp. 107–11.

Knight, F. (1921) *Risk, Uncertainty and Profit*. Boston: Houghton Mifflin.
Kochanek, S.A. (1996) 'Liberalisation and Business Lobbying in India', *Journal of Commonwealth and Comparative Politics*, vol. 34 (November), pp. 155–73.
Leff, N.H. (1979) 'Entrepreneurship and Economic Development: The Problem Revisited', *Journal of Economic Literature*, vol. 17 (March), pp. 46–64.
Lokanathan, P.S. (1935) *Industrial Organisation in India*. London: Palgrave Macmillan.
Marshall, A. (1890) *Principles of Economics*. London: Macmillan Palgrave.
McClelland, D. (1961) *The Achieving Society*. Princeton, NJ: D. Van Nostrand.
McClelland, D.C. and Winter, D.G. (1971) *Motivating Economic Achievement*. New York: Free Press.
Nafziger, E.W. (1971) 'Indian Entrepreneurship: A Survey', in P. Kilby (ed.), *Entrepreneurship and Economic Development*. New York: The Free Press.
Nafziger, W. (1975) 'Class, Caste, and Community of South Indian Industrialists', *Journal of Development Studies*, vol. 11 (January), pp. 131–48.
Nafziger, E.W. and Rao, R.S. (1996–97) 'Industrial Entrepreneurship and Innovation under Licensing and Liberalization in India', *Indian Economic Journal*, vol. 44 (October–December), pp. 90–101.
Pedersen, J.D. (2000) 'Explaining Economic Liberalization in India: State and Society Perspectives', *World Development*, vol. 28 (February), pp. 265–82.
Schumpeter, J.A. (1934) *The Theory of Economic Development*. Cambridge, Mass.: Harvard University Press.
Schumpeter, J.A. (1939) *Business Cycles*. New York: McGraw Hill.
Singh, S. (1985) 'Relevance of Social Factors in Entrepreneurial Growth', *Journal of Sociological Studies*, vol. 4 (January), pp. 72–85.
Sinha, D. (1995) 'Caste as an Economic Institution in India: An Interdisciplinary Study', *Asian Economies*, vol. 24 (March), pp. 27–45.
Sinha, T. and Sinha, D. (1994) 'Opportunities in India: Consequences of Liberalization and Globalization', *Journal of International Marketing*, vol. 2(2), pp. 3–10.
Sridharan, E. (1993) 'Economic Liberalisation and India's Political Economy: Towards a Paradigm of Synthesis', *Journal of Commonwealth and Comparative Politics*, vol. 31 (November), pp. 1–31.

9
Angels, Venture Capital and Entrepreneurship

C. Gopinath and Navendu Vasavada

Economists have traditionally considered four factors of production: land, labour, capital and entrepreneurship. The entrepreneur was the idea generator and the risk-taker who combined the other three factors in creating an enterprise. Apart from own-funds, the entrepreneur's ability to raise funds from outside is crucial in bringing the idea to fruition and this is where the role of the venture capitalist (VC) comes in. The specialized nature of the activity of funding entrepreneurs has resulted in three sources of funding:

1. The 'angel' who is a wealthy individual who invests in amounts that provide the seed money for a venture and helps the entrepreneur to initiate activities and be able to approach larger institutions.
2. The venture capital firm that operates with own-funds and also serves as a conduit for other individuals and funds that seek high returns.
3. Venture capital operations of larger banks and financial institutions.

At each stage, the angel and/or the VC is constantly seeking to diversify their risk of exposure by getting the firm ready to seek market funding through an initial public offering (IPO).

In this chapter we first describe the role of VCs in the USA as an example of how the function operates in an advanced capital market environment. We then describe the case of The IndUS Entrepreneurs (TiE) as an organization that seeks to serve entrepreneurs by combining different aspects of the entrepreneurship process including venture funding. We conclude by calling for a rethinking in the policy approach towards venture capital and entrepreneurship in India in order to build a supportive infrastructure.

The traditional large corporate sector

Entrepreneurship in the large corporate sector takes place in an environment of plenty. Capital for new projects and innovation is easily accessible in the US corporate sector from both internal sources and a well-developed capital market. The internal corporate R&D budgets of large US companies fund innovation and initiatives. A large company's R&D expenditure is absorbed into its internal cost structure, and corporate R&D is the basis of large companies gaining advantage in the technology and pharmaceutical sectors. Accompanying large R&D budgets are large organizations with research managers and scientist hierarchies. While large-scale research laboratories are able to create profitable innovations, they come at a slow pace, and are subject to low-cost predatory R&D by upstarts.

Window of entry for new startups

Over the last four decades, only three areas of competitive entry by upstarts have come to the forefront. The first is computer hardware and software, the second biotechnology and the third internet-related e-commerce. The first two have established track records of successful entry and victory over competing large corporations. It is premature to judge whether the third new sector of internet-related e-commerce will be sustained and persistent.

Biotechnology has relatively fewer cases of success when compared to telecommunications and computers. That is because the gestation period of biotechnology projects is very long and the success rate very low, making it less attractive for investors with little capital and short horizons. Only well-funded biotechnology startups funded by deep pocketed investors with very long time horizons have been able to surface with spectacular offerings. In contrast, the area of computer hardware and software has produced a virtual cornucopia of successful startups.

The new startup technology ventures do not have the deep pockets to navigate through uncharted territories of fundamental research. This naturally falls under the realm of deep-pocketed established corporations with well-funded research laboratories.[1] The nimbleness of new startups in design and marketing give them an edge over the corporate giants, who leave large spaces in the marketplace open to startup companies.

Size of the US venture capital sector

One of the most followed surveys of the US venture capital investment is published by PriceWaterhouseCoopers jointly with the Reuters affiliate

VentureOne.[2] This survey tracks the number of deals, size of deals and breakdown by specific sub-sector. The number of deals steadily rose from 1174 in 1995 (valued at $6.3 billion) to 4107 in 2000 (valued at $68.6 billion). Examining the breakdown of these numbers by specific sectors, it is not surprising that the top five sectors that received the funding in 2000 were Consumer/Business services (dot.coms), communications (including internet), software, information services and semiconductors. Various medical-related sectors including biotechnology follow in the sixth position onwards.

The funding of $68 billion in 2000 is unlikely to be sustained.[3] Consider a perpetual fixed annuity of $68 billion per year: at current interest rates, this annuity is worth at least $1.4 trillion, so that if investors were funding the venture capital segment at a sustained constant rate of $68 billion per year they have to forego a secure bond portfolio worth $1.4 trillion. This is a significant absolute amount of secure investment to forego. Further, this amount of $1.4 trillion foregone is significant in comparison to the entire US bond and stockmarkets, which currently have capitalization of $9 trillion and $10 trillion respectively.

It is more likely that the sustained average annual funding in the US venture capital segment would be similar to $10–15 billion per year, the approximate level that existed in the period 1996–98. This is one-fourth to one-sixth of the funding of $68 billion in 2000.

The arrival of 'Angels'

Conditioned by the success of computer ventures, a culture of venture capitalism arose largely in California, with capital largely provided by institutional investors. A community of individuals also known as 'Angels' emerged who absorb the tremendous risk of losing their entire small investment in taking a concept to a credible business plan to present before a VC firm.

Organized VC firms expect to see a well-composed project funding proposal, which has transformed itself from the state of a pure idea to proof of concept, a management team which is capable of bringing the product to market, and a chief scientist/technologist who implements the concept. Inventors and originators of ideas typically lack the modest capital to meet the smallest of startup expenses and salaries for associates and employees.

While some Angels are wealthy passive investors, most are typically active investors who acquired their wealth from previous spectacular successes, as an entrepreneur or an early employee in a successful startup company. Such Angels recognized that their association with previously successful venture ideas was a credible signal to VC firms of prescreening and soundness of the idea. Active angels come with knowledge and experience in screening ideas and transforming them into a credible business acceptable by VCs for funding. They signal their own willingness to take risk and lose their entire investment, and display their track record of previous successful projects to VCs as an indicator of likely success of the proposed venture.

Passive angels seek to earn large returns on their initial investment, and reflect their risk tolerance to total losses of their investment. They are largely uninterested in actively managing or participating in the new venture, and usually participate in syndicates and consortia to spread their risk across many venture ideas while ensuring that some active angel who is taking about the same risk is actively managing the relationship with the concept originator.

Active angels usually specialize in very narrow areas, especially those in which they have themselves been successful. Thus, they are able to evaluate, based on personal knowledge and experience, the market and technical viability of an idea. This is of value to the VC firm too. By contributing funds to an entrepreneur and being associated with the project, angels lend something of greater value than the funds – their credibility and expertise. A VC firm is more likely to entertain a request for funding from an entrepreneur who has a well-known and successful angel associated with it.

Dividing the expected profits with angels

The key question at all stages of a venture project is: what is its pre-money valuation? This venture capital acronym refers to the hypothetical market value of the project prior to obtaining funding from an Angel or a VC. Angel syndicates seek to provide startup cash financing on terms such that the hypothetical valuation times their stake equals the cash that they provide. Thus, $1 million start-up capital for a 10 per cent stake implies a 'pre-money valuation' of $10 million.

Angels, typically, seek terms that sometimes cause alarm to the idea founders. During the wave of venture activity in 1998 and 1999, idea originators were able to get angel financing at the rate of $1 million for a 15 per cent stake ($7 million pre-money valuation). This is the lower end of angel terms from the idea originator's point of view. Angel terms can be as much as a 40 per cent stake, or even more than 50 per cent for $1 million financing ($2.5 million valuation).

Clearly, the angel cash infusion alone is usually insufficient to get the project to market. It starts the project and pays for the initial zero-order costs of scaling up an idea to a credible business for consideration by VCs for further financing. The angel and idea originator typically expect VCs to come to terms with them for a pre-money valuation of $40–100 million. The VC finances the next round of capital expenditure, say $10 million, at a scaled-up $50 million pre-money valuation. The VC thus acquires a stake of 20 per cent in the project, by diluting the stakes of the angel and the idea originator. The angel owns 15 per cent of the remaining 80 per cent of stake, which is now 12 per cent of the total, and the idea originator owns 85 per cent of the remaining 80 per cent stake, which is now 68 per cent of the total.

As the project attracts more VC attention and is ready for more capital infusion to bring it up to operating scale, it systematically raises its pre-money valuation at each successive round of VC financing, such that just prior to IPO, the idea originator has 30 per cent of the stake, the angel about 5 per cent stake, and the VCs hold the remaining 65 per cent. This was the typical experience of mega-IPOs in 1999.[4]

The angel ends up with about 5 per cent or less of the stake through successive rounds of dilution, from the stage of investment to the stage of completion and exit through IPO. This is a large reward in a mega-IPO. If the typical IPO market value is $1 billion, the angel gets $50 million, for having put up $1 million at the inception. Thus, the order of magnitude of the angel's investment gain is about 50-fold, contingent upon success of the project and its acceptance in the marketplace.

Convergence from hypothetical valuation to real market valuation

The pre-money valuation at successive stages of the venture is elevated to converge to IPO value of the new project. This was a very profitable exit strategy for angels and VCs alike, to the extent that they were permitted to sell some part of their stake within a short period from the time of the initial public offering. The decline of the NASDAQ by over 60 per cent during 2000–01 has brought a harsh correction to lofty IPO values. It is now clear that many projects were financed in 1999 and 2000 and released to the IPO market at valuations that were unrealistic and did not converge with the market valuation by March 2001.

To the extent that the hypothetical pre-money valuation does not converge to the market valuation in the first few years of the life of the startup company, there are deep losses borne by angels, VCs and market investors who acquire their stakes shortly after its IPO.

The NASDAQ market correction of 2001 introduced a stark reality check on the pipeline of new ideas. The hypothetical pre-money valuation is subject to very stringent examination of assumptions and conjectures by angels and VCs alike. The flow of such ideas advancing to the IPO stage in the computer and internet sectors would be diminished, because angels and VCs have to exercise more risk-aversion and prudence in screening projects. On the other hand, we are likely to see an increase in the pipeline of biotechnology projects since their prospects are based on hard science and are unaffected by the market's concerns about IPO valuations of computer and internet projects.

Risk and expected return

We have estimated the payoff to the angel, as much as 50-fold, within a scenario of success. What are the odds of success? We believe that they are very remote, less that 1/50 to 1/100, so that the expected return may actually be negative except in the very early phases of market expansion. If a consortium of angels were to invest in 100 projects, at the rate of $1 million in each project, they need a success hit-rate of at least two projects producing 50-fold gains just to break even. A higher hit-rate was possible only in the very early stages of the Internet wave of the late 1990s. The angel bears the enormous risk of project abandonment due to the inability of the project to attract VC financing. Even if the idea is sound, the project is a complete loss to the Angel if VCs reject it.

VCs also take on deep risks but they appear to be taking less risk relative to angels. During 1998–99, VCs had put up as much as $65 million to acquire a 65 per cent stake in a project at $100 million pre-money valuation. If the project fails to reach the IPO market and acquire a market valuation of $1 billion, the VC has lost their entire $65 million. However, it appears that the odds of market success at this stage of the venture are higher than 1/50 to 1/100, which the angel faced. It was closer to 1/10 or 1/20 for a VC prior to the NASDAQ decline in late 2000. Thus, at a $1 billion market valuation, the VC gets a payoff of 10 times. The VC must have a success hit-rate of 1 in 10, translating to success odds of 1/10, to break-even in terms of expected value. In contrast, the angels face a much lower success hit-rate, and despite lower investment, seem to face a much lower expected value as compared to VCs.

Due to their deep pockets, VCs can set up organizations and diversify their investment across a sufficiently large number of ventures, whilst angels are unable to diversify to the same extent. An angel who typically faces odds of 1/100 needs to invest in 100 projects to pay for losses. This is a practical impossibility for angels due to limitation of time, knowledge base and capital. We have seen that angels take much higher risks (as defined by expected value) compared to VCs, but it is not in the interest of VCs to eliminate them as middlemen and directly cut deals with idea originators. The screening, signaling and management preparation role of the angels, combined with the extremely high risk taken by angels, is very attractive to VCs.

TiE as a unique entrepreneurship model

Our discussions till now have focused on the financing and risk-taking roles of angels and VCs. We now introduce The IndUS Entrepreneurs, known better by its acronym, TiE, which permits appreciation of the larger context of entrepreneurship development and how funding considerations are woven into it. TiE is an organization that likes to be known as 'A non-profit global network of entrepreneurs and professionals, established to foster entrepreneurship and nurture entrepreneurs'.[5]

TiE received a lot of publicity due to several cases of success amongst its members who, in the heydays of market euphoria, achieved very high capitalizations for their companies. TiE members have created businesses with a market value of $75 billion since 1992, according to the group's estimates.[6] Even if this were deflated by 80 per cent, according to the decline experienced by unseasoned technology companies in 2000–01, the surviving market value of $15 billion is still a respectable sum. Although TiE began with its roots in the expatriate South Asian community, it now seeks to serve entrepreneurs in general. While the majority of TiE's members are from the Indus region (India, Pakistan, Bangladesh, Nepal and Sri Lanka), there are non-Asian members too. The organization's materials declare that while it originated in the Indus culture, it now offers membership to everyone regardless of ethnicity and religious affiliation.

Indians, in general, have made a significant impact in founding new businesses in the computing technology and telecommunication fields. Indians founded 9 per cent of the startups in Silicon Valley between 1995 and 1998, up from 3 per cent between 1980 and 1984, according to a study by the Public Policy Institute of California.[7] The study also found that Indian-led high-tech companies in Silicon Valley generated $3.5 billion in sales and employed more than 16 500 people in 1998.

Another study showed that Indian entrepreneurs headed 774 of the 11 443 high-tech firms started in the valley since 1980.[8]

TiE began in 1992 when a group of Silicon Valley executives of Indian origin got together to meet with a visiting Indian government official and decided to continue meeting for the purpose of mentoring and networking. In 1994, the group grew into a formal organization with the objective of building a network to help South Asians succeed in Silicon Valley. South Asians were a significant force in the US technology sector long before the Internet boom and the emergence of TiE. In the 1970s, for example, several Indian engineers were doing important work in AT&T, IBM and Xerox. However, they frequently found that they came across a glass ceiling that blocked their ascent into upper levels of management. A similar barrier seemed to present itself when they sought assistance to pursue entrepreneurial dreams. Thus, TiE saw itself as a way to smooth the process for immigrants with ambitions.

The visible proportion of Indian technologist entrepreneurs can be traced to a steady growth in the proportion of US-educated Indian engineer technologists. Beginning in the 1970s, there has been a secular decline in the proportion of graduate electrical engineers who are US-citizen Americans. The most recent Survey of Earned Doctorates by the US Government[9] records that 6052 doctorates in engineering were awarded by US universities in 1997, compared to 2604 in 1967. However, the proportion of US citizens receiving PhDs in engineering declined from 70 per cent in 1967 to 44 per cent in 1997. The breakdown of PhD engineering recipients of foreign origin in 1967 is not available. In 1977, of the 56 per cent of total US PhD engineering degrees awarded to recipients of foreign origin, China and India accounted for 10 per cent each, followed by Korea at 7 per cent and Taiwan at 5 per cent. All other countries had negligible proportions, well under 1 per cent, including Japan and the UK.

TiE has two categories of members, regular Members and Charter Members. Charter Members, who are by invitation only, embody one aspect of its mission that TiE considers very important. They are 'successful veteran entrepreneurs, corporate executives and senior professionals who have reached a stage in their professional life when they are ready, willing and able to contribute to fellow members'.[10] Kanwal Rekhi, a pioneer in TiE, is quoted as saying, 'One of the values we expound at TiE is the notion of payback'.[11] At the time of writing, TiE has 31 chapters spread across the USA (17), India (7), Pakistan (2), Canada (2), Singapore (1), Dubai (1) and the UK (1). It has plans for continued rapid growth.

The activities of TiE chapters are built around events such as an inspirational talk by a successful entrepreneur, or workshops and seminars on aspects of entrepreneurship and attracting capital. Such an event provides the setting for a social period of intense networking before and after the speech or workshop. In some events, there is an open microphone period, part of an 'Angel Forum' in which anyone who wants the chance gets 90 seconds to pitch his or her ideas.

TiE, through a process of evolution, seeks to serve several purposes:

- It provides the social atmosphere wherein budding entrepreneurs can interact with and recognize that others have gone through the trials they face and have emerged successful. Thus, it provides the inspiration and motivation to sustain effort. In the words of one entrepreneur, 'One of the key things we've learned in entrepreneurship is that there's nothing that encourages individuals more than getting to meet someone who has become successful, and say, "If he did it, I can do it"'.[12]
- In a more intimate manner, entrepreneurs can find mentors to seek advice and opportunity. This is done by individuals seeking out one of the other members and building a one-on-one relationship that has been likened to the 'guru–shishya parampara' or the teacher–student relationship in Indian tradition.
- It provides an avenue for angels and venture capitalists to find opportunities to invest in.

While 'networking' has always been considered a part of the social life of business, its role in entrepreneurship is increasingly receiving attention.[13] For new ventures to be successful, it is very important to build a network of supportive relationships. Apart from being potential sources of funds, members of the network provide non-financial resources, too, such as introduction to suppliers, customers and professional advice.

As an organization, TiE does not invest but its members do. Creating an environment for funding ventures is an important part of the activities of the organization. A membership survey in 2000[14] showed that 71 per cent of the members were looking to interact with Charter Members with the objective of venture capital funding, and 70 per cent did it for networking. Again, 48 per cent of members reported that their interest in workshops was for one-on-one interaction with VCs; 97 per cent of members gave 'networking' as the reason for joining the organization and 71 per cent mentioned 'mentoring'.

The mentoring/funding process becomes self-sustaining as new and successful entrepreneurs are created who in turn become mentors, providing expertise and seed money for other enterprises. For instance, in 1995, Rekhi invested $250 000 and his expertise into Exodus Communications, a startup by K.B. Chandrasekhar. He also joined the board and helped the firm set a direction. Rekhi's stake multiplied several-fold with a quick rise in the value of Exodus' stock. Leveraging the success, Chandrasekhar became an investor himself, backing several companies and serving on the boards of some.

In 1996, Suhas Patil, founder of Cirrus Logic, invested $2.8 million in RightWorks, a firm founded by Vani Kola, and got back more than $12 million. Kola has since left RightWorks, which sells software to companies doing business on the Internet, and has started mentoring and investing in aspiring entrepreneurs.[15] These examples show the progression of angel activities over successive generations of entrepreneurship, with successful entrepreneurs turning into angels and mentors themselves.

There are several other organizations that are associated with ethnic efforts to build a network and foster entrepreneurship.[16] Yet, by combining networking, funding and mentoring, the benefits that TiE provides appeal to a growing band of potential entrepreneurs.

TiE organized itself in a timely and spontaneous manner to position itself to meaningfully participate in the period until 1999 when the NASDAQ reached its peak. It focused on the core strengths of its constituents: entrepreneurship largely in areas related to computers and the internet; most of these projects were based on low capital and short gestation. It is very unlikely that global equity markets will provide entrepreneurs with computer and internet-based projects with a quick exit strategy. The future success of TiE will be based on how its members are able to absorb risks on long-gestation projects without a quick IPO-based exit, and how they adapt to activities in areas other than computers and the internet, particularly in biotechnology. Yet, the continued growth of the organization shows that it fulfils a felt need.

Entrepreneurship and the VC market in India

Entrepreneurship in India has seen a resurgence of interest with market reforms and liberalization. More recently, interest has been fuelled in the area of computer software, which is seen as requiring less capital investment and government permissions.

Entrepreneurship is often considered synonymously with small business. In an environment where the government is seen as paternalistic and protecting weaker sections of society, policies towards entrepreneurship and small business have been weighted towards protection. Thus, there are reservations of items for small business so that they do not fear entry and competition from large businesses. One consequence of this has been to create disincentives for a small business to grow large as it would result in loss of protection. It also fosters a high cost structure (in the absence of competitive pressure and scale economies). Apart from product reservations for small businesses, the policy of the central government has been to provide favourable tax policies such as excise tax exemptions.

The underdeveloped nature of the capital market has meant that a typical entrepreneur has had to largely rely on private placement (with family and friends), loans from financial institutions, and IPOs. VC funding has been mostly through government-owned institutions such as the Industrial Development Bank of India, and state financial/industrial development corporations. The private VC industry in India is still in the rudimentary stages. Government-owned institutions are interested in safeguarding and recovering their investment while driven by social considerations of promoting industrial development. Their lack of a hard-nosed profit-oriented approach tends to weak selection of entrepreneurs to promote, and a financial/lending focus rather than a business focus.

As a major policy shift, the government must see the process of entrepreneurship as a dynamic activity separate from the concerns of small businesses in general. As the TiE case indicates, the venture capital industry must not be seen merely as provider of funds but as a key player in an entrepreneurship process. Through the process of providing funds, they identify business ideas, mentor inexperienced managers, develop business leadership, and even attract talented employees for the new venture. Without their contribution, several ideas and the opportunities they provide would fizzle and die. And there is a need for the right set of government policies to facilitate this. Figure 9.1 tries to capture the spirit of this interaction in terms of the roles to be played.

The Government of India's policy towards venture capital has recently been going through a process of evolution. While the government is keenly aware of the need to facilitate the process of new-venture creation, a comprehensive policy framework does not appear to be in place and those actions that are being initiated are mostly in response to external demands. In general, it appears that the government has taken the approach of following fiscal controls rather than physical

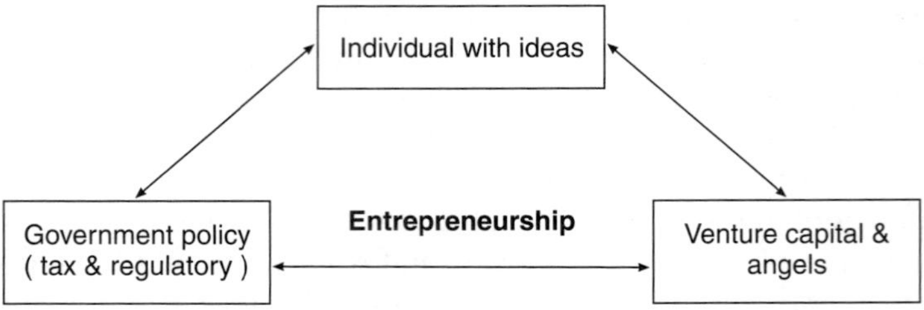

Figure 9.1 Players in the process of entrepreneurship

controls on venture capital activities, which is certainly in the right direction.

Much of the impetus for restructuring the regulatory framework appears to have come from successful non-resident Indians in Silicon Valley, such as the members of TiE. With their experience and wealth, they are able to seek policy changes that would facilitate their investment in India, directly or through venture capital funds.[17] Since 1996, the government has been attempting various reforms affecting the VC industry. These are in the areas of the regulatory framework (rationalizing the regulatory process, reducing the number of regulatory agencies overseeing the VC industry, eliminating overlap), taxation (eliminating differentials between overseas and domestic funds, treatment on par with mutual funds), raising funds (sectoral restrictions, exit rules), and so on.

These reforms are often piecemeal and seemingly random.[18] Random regulations and frequent revisions are often in response to demands. What is necessary is to look at the role VC plays in the broader context of entrepreneurship and devise a new vision that breaks away from seeing the small sector as one to be protected to viewing entrepreneurship separately as a dynamic new venture and wealth-creating activity. A consistent policy that will stay in place for a sufficient period of time would allow VCs to plan their activities. The nature of VC operation is very different from that of a portfolio investor and this should be reflected in the policy.

Environmental conditions that support venture capital formation

There needs to be an appropriate set of macro economic, political, regulatory, social and other factors in existence that would allow VCs and angels to pursue their interests, while at the same time harnessing their efforts to foster capital formation and job-creation in the country. In the US case these factors include:

1 A deep capital market in an active mature market economy that fosters creativity and individuality (note that Japan has a deep capital market but not the same opportunities for individual creativity to flourish).
2 A system of law and enforcement mechanisms which allows innovators to seek protection of their intellectual property at a low cost, and provides a safe harbour and recourse from fraud and embezzlement. It is much easier to siphon funds out of a company in an emerging market than in the USA, where penalties for such actions are very harsh and enforcement is virtually certain.
3 Favourable tax treatment of profits as long-term capital gains. Such profits, taxed at about 50 per cent of the rate on ordinary income, compensate investors for the increased risks and make venture capital an attractive investment.
4 Regulations allowing limited partnership enterprises, which enables the investor to obtain tax advantages on his investment. Other structural factors include the need for full disclosure, and rules requiring sufficient shares to be issued to ensure liquidity, apart from policing market abnormalities.[19]

Conclusion

Successful entrepreneurship in a society emerges through a combination of several factors. Our focus has been on the role of angels and private venture capital, which is only one but an important aspect requiring special attention. In the past, policy-makers in India have viewed venture capital as a source of funding alone. With several other demands on its funds, the Government of India has strong reasons for creating the right environment to allow angels and private VCs to flourish. It would also be in keeping with the government's general reform policy of reframing its role in industrial development from a direct to an indirect one. Indigenous company growth is an important component of a long-term economic development strategy and new venture creation is a critical part of that. It is time to consider a radical reorientation of thinking. A proactive approach towards fostering the entrepreneur's risk-taking and innovation through venture capital funding would contribute more to society's wealth and employment opportunity than one of protection.

Notes

1 Research labs such as Bell Labs of Lucent, and Xerox's Palo Alto Research Centre have this luxury.
2 This report is available on the Price Waterhouse Coopers website at http://www.pwcmoneytree.com/. However, this website could be in a state of flux, and the survey report may not be available as a permanent reference.
3 For the full year 2001, it is estimated at $30 billion.
4 For example, firms such as Juniper and Sycamore fell into this pattern when the US stockmarket reacted favourably to the high-technology ventures.
5 www.tie.org
6 J. Guynn, 'Silicon Valley Entrepreneur Establishes a Powerful Networking Group', *Contra Costa Times*, 17 July 2000.
7 *Ibid.*
8 AnnaLee Saxenian (1999) *Silicon Valley's New Immigrant Entrepreneurs*, Public Policy Institute of California, San Francisco, CA., p. 23.
9 'Summary Report 1997, Doctoral Recipients from US Universities' by the National Science Foundation and other US agencies. This report is currently available on the internet at http://www.nsf.gov/sbe/srs/sengdr/summ97/start.htm
10 www.tie.org
11 J. Guynn, *op. cit.*
12 Safi Quereshey, quoted in D. Kesmodel, 'South Asian Entrepreneurs and Mentors Form a Tie that Binds', *Los Angeles Times*, 3 July 2000, p. 1. Qureshey, the President of TiE Southern California, is of Pakistan origin and achieved prominence in the 1980s as cofounder and Chief Executive of Irvine-based personal computer maker AST Research Inc.
13 L. Steier, and R. Greenwood (2000) 'Entrepreneurship and the Evolution of Angel Financial Networks', *Organization Studies*, vol. 21(1), pp. 163–92.
14 www.tie.org/survey-results-2000.htm
15 B. Stocking (2000) 'California Group Aids Indian', in 'South Asian Technology Entrepreneurs', *San Jose Mercury News*, 27 November.
16 An earlier ethnic-based organization that has made an impact in the hospitality industry is the Asian American Hotel Owners Association, which began in 1989. The AAHOA also includes mentoring and training as part of its activities. However, angel/funding activities within this community are not well-documented.
17 A most direct recognition of this influence can be seen in the appointment in July 1999 by the Securities and Exchange Board of India of Mr K.B. Chandrasekhar, Chairman of Exodus Communications, to head a Committee to make recommendations for government consideration. SEBI is said to be looking to model a framework based on the experience of the USA and Israel. See (a) K. Merchant, 'Venture Capital: Struggle to Bridge the Management Gap', *Financial Times*, 13 September 2000, p. 8, and (b) L. Joseph and R. Pinto, 'Waiting to Take Off', *Business India*, 7–20, February, pp. 85–8.
18 Consider the following example. In September 2000, under the initiative of the Finance Ministry, SEBI's guidelines for VCs included: (a) mutual funds may invest in local VC funds, (b) open-ended mutual funds may invest up to 5% and close-ended funds up to 10% of their funds available for investment

in venture capital funds. (c) VC funds may invest up to 75% of their funds in unlisted equities, and (d) up to 25% of their funds in initial public offerings with a 1 year lock-up period or in debt instruments of companies in which they already have an equity investment (AFX News Limited, 'SEBI Permits Mutual Funds to Invest in Local Venture Capital Funds', 14 September 2000). Not long after, the Finance Minister changed the government's position and said VCs need not exit the companies even after one year of making an initial public offer, and yet will receive tax breaks on any gains.

19 For further discussion, see G. Kozmetsky, M.D. Gill and R.W. Smilor (1985) *Financing and Managing Fast-Growth Companies*, Lexington, Mass.: Lexington Books.

10

Business Strategy and Corporate Restructuring in India: The Mutually Reinforcing Impacts of Economic Liberalization and Technology

*Raj Aggarwal**

Introduction

The Indian economy has been on a deregulatory path since at least the early 1990s and the process of economic deregulation and external opening seems to be accelerating. These changes are forcing Indian business to become domestically and globally competitive and it faces major changes in corporate structure and management culture. There are a number of trends that are accelerating this process. First, deregulation and global opening of the Indian economy is firmly established and unlikely to reverse regardless of the type of government in power (though its speed may vary). Second, Internet-based technologies are rapidly being implemented in India. These technologies speed up business processes, eliminate most impacts of distance, and generally reduce transactions costs.

Not only is each of these two forces individually very powerful, they are mutually reinforcing. As deregulation and global opening of the economy forces many Indian companies to become more efficient and globally competitive, they will inevitably turn to the new Internet-based technologies. With technology, Indian firms can also start to exploit foreign markets. Indeed, technology enables globalization and globalization makes technology more profitable (Aggarwal, 1999). In this

---

* The author is grateful to his colleagues, J. Ryans, and A. Sinha for useful comments, but remains solely responsible for the contents.

process, the first few successful Indian companies will serve as centres of excellence, as role models, and as sources of management know-how for other Indian firms. As the use of technology spreads to second and third-tier firms, it will fuel further demand for technology and improve the efficiency of ever-larger proportions of the Indian economy. These developments in the institutional environment of business are likely to provide an important boost to India's long-run economic growth rate.[1]

However, these mutually reinforcing positive developments also mean a number of associated changes in the business environment, some of which are likely to be very disruptive. Most large Indian firms will need to be restructured and will be forced to define and focus on their core competencies. Defining and focusing on core competencies is often a difficult and slow process. As some firms become more efficient, they are likely to drive out firms that are slow to change and remain inefficient too long. Corporate bankruptcies, especially large ones, always have a tendency to get messy, frequently with political implications. Indian business certainly faces a period of accelerated change, but it will also be a period with many opportunities.

Deregulation, technology and Indian economic growth

India is a large and strategically important democratic country with significant nuclear and aerospace capabilities. It has the second largest population in the world and is a major military and economic power in Asia. As measured by nominal GDP, India is the eleventh largest economy. But if economic size is measured by purchasing power, that is based on global prices for what people buy, then according to World Bank measurements India is actually the fourth largest economy in the World, just behind the USA, China and Japan (*Economist*, 12 May 2001, p. 110). Further, in the last decade, according to the 2000 *Annual Report* of the Bank for International Settlements, India's real economic growth rate has exceeded (with very few but notable exceptions) the growth rates of most other economies including fast-growing Asian economies such as Taiwan. What challenges does India face as it strives to maintain this superior economic performance in the next decade? Clearly, one of the primary challenges is to maintain the momentum of economic deregulation.

Development dilemma and deregulation

While there are already some world-class businesses based in India, most Indian businesses have until recently been better at dealing with the bureaucracy than with competitors. Even though there has been some domestic competition in most industries, many Indian companies basically faced a seller's market – indeed the intention of the bureaucracy had generally been to ensure that overproduction was minimal and unintentional. This is now changing, and the deregulation of Indian business means that it must now face increased competition not only from other domestic companies, but also from foreign companies. What forces are driving this process of deregulation and is it reversible?

A major social force that impacts on developmental policies in India is democracy. India is not only the world's largest democracy, it has the second largest population with 70 per cent living in rural areas. India also has large numbers of distinct linguistic and ethnic groups, there is much poverty, and large regional variations in income and wealth (e.g. Budhwar, 2001). In spite of these variations, democracy is well-established as Indian cultural values seem consistent with the requirements of democracy. In contrast with other countries at low levels of economic development, the level of interpersonal trust is high in India exceeding the level found in many developed societies, and India also ranks with the developed countries in measures of freedom and opportunities for self-expression (Inglehart, 2000). Reflecting these values, India has been a secular democracy since its independence from Britain in 1947. While rare among developing countries, Indian democratic institutions seem robust having withstood many shocks over the last half century. Thus, it is reasonable to assume that Indian policies for economic development, including the process of deregulation, are consistent with, and can depend on, a stable democratic political structure.

It is useful to examine recent history briefly in order to understand the nature of economic deregulation in India and to understand how and why Indian business has been in a bureaucratic straightjacket until recently. At independence from Britain in 1947, almost all of the Indian leadership had been trained in England by (well-meaning) Fabian Socialists. These leaders were responsible for setting up state machinery designed for the bureaucratic control of the Indian economy – with this bureaucratic machinery remaining essentially unchanged until the 1980s.

One of the reasons why the Indian bureaucracy had been so draconian is that until 1990 it was felt (incorrectly) that such bureaucratic control was the only way in a democracy to keep in check and defuse poverty and inequitable wealth-distribution-driven political unrest. The high population growth rate requires ever-increasing resources for food and shelter, and as the Indian government tries to contain the number of people living in poverty, it must attempt to redistribute some wealth through taxes and regulatory means. Unfortunately, as these tax and regulatory burdens increase, economic growth rates inevitably drop. So any democratic Indian government faces the same basic developmental dilemma; that is, how much economic growth should it give up for a more even distribution?

In the past the government had put too much emphasis on equality at the detriment of economic growth. Now there is a realization that the higher levels of economic growth that are possible with deregulation can mean that everybody, including the poor, may be better off. One reason is that it is well-documented that wealthier people have fewer children reducing the pressure on resources resulting from rapid population growth. In addition to this effect of economic growth, the aggressive population-planning policies of the Indian government that are helping the country reduce its rate of population growth are easier to fund when there is higher economic growth. So, it is clear that higher economic growth and lower population growth are mutually reinforcing. Thus, it is likely that the population problem in India will be contained sooner than most people think, and is not likely to derail the deregulatory process.

The current process of economic deregulation started in the early 1990s. At that time, India faced a number of pressures to deregulate its economy. First, economic growth had increased to about 5 per cent from the anaemic 3–3.5 per cent annual rate of the prior decades, but was now stalled. Most importantly, India had borrowed heavily to finance its growth in the 1980s and now faced critical balance-of-payments crises with the prospect of international default for the first time since independence in 1947. Second, the Berlin wall and many socialist and communist states had fallen in 1989 and there was a strong international movement towards market-driven economies. Among India's neighbours, the Soviet Union was falling apart and China was having much success with the economic deregulation it had started in the early 1980s. Third, even big business in India that had long accepted protected markets in exchange for draconian bureaucracy, now wanted change and opportunities for growth (e.g. Percy, 1992). This combination of internal and external pressure resulted in the first wave of economic deregulation in India in the early 1990s.

This first wave of deregulation freed up enough of the large amounts of pent-up demand among Indian consumers that it was considered a major success by government, industry, and consumers (e.g. Joshi and Little, 1996). This initial success paved the road for subsequent deregulatory moves, which have also been mostly successful. Consequently, by now the major political parties, business and industry, and consumer groups, are all committed to the deregulation and the gradual global opening of the Indian economy (e.g. Jishnu *et al.*, 1999). Indeed, India is not only committed to membership in the World Trade Organization (WTO), but also to a leadership role in the organization. However, as can be expected in any democratic country, there continues to be much public debate in India about the nature, sequence, and speed of the deregulatory process, but there is now little doubt about the need to deregulate the Indian economy.

Computer technology and Indian economic growth

India has many resources that account for the extraordinary success of its computer and software industry. India has a superb educational system and graduates one of the largest numbers of scientists and more than a quarter million engineers annually; and many Indian engineering institutes are truly world-class.[2] In addition, intellectual achievements are valued highly in Indian culture. English is the medium of instruction for much of higher education and, with the revolution in electronic communications, Indian knowledge workers that earn much less than their counterparts in the developed world can, and have easily, become part of the global economy. Finally, many non-resident Indians (NRIs) are investing and moving back to India, actively participating in transferring capital, technology, and management know-how from western countries to India.

As in the rest of the world, internet-based technology is starting to change the nature of business in India and many Indian companies have become major, globally-competitive, world-class suppliers of software and other computer services to the rest of the world (e.g. Das, 2001). Many of these companies are located in close proximity to each other in centres like Bangalore, Hyderabad, Delhi and Mumbai, giving rise to the advantages of agglomeration and specialization in developing world-class firms (e.g. Porter, 1990). This rise of India as a source of technology for the rest of the world has important implications for the growth of the Indian economy.

Indeed, it has been estimated that more than one-third of Fortune-500 companies already have computer software development operations in India. And this proportion is increasing as more US, European and Asian companies are starting to follow this initial group. Further, because of its large English-speaking population, India is becoming the back office and customer service division for large numbers of US and European companies who have been setting up call centres and back-office operations in India. These activities are creating jobs and adding to domestic economic growth.

In addition, the growth of any industry creates a number of ripple effects. The rapid growth of the Indian software and computer services industry has also led to increased rates of growth in other (supplier) industries, for example in industries such as telecommunications, computer hardware and other products and services consumed by the software industry.

The growth of a world-class industry like software also means that there is much technology transfer to other Indian industries, and technology has widely been considered a prime source of economic growth (e.g. Helpman, 1998; Mokyr, 1990). It is clear that global levels of excellence in technology and management practices are being achieved in the Indian software industry. As management know-how and technology in one industry inevitably gets transferred to other industries, these high standards trickle down to other industries, making other Indian products more desirable and competitive globally.

The growth of the software industry also helps the Indian economy as it is becoming a major source of export earnings. In addition, Indian software companies have been able to raise substantial amounts of money in global financial markets, such as New York, London and other financial centres. Table 10.1 lists the largest Indian companies that trade as depository receipts in the United States. As this table shows, the list is currently dominated by software companies. In addition to the New York Stock Exchange and the NASDAQ in the USA, securities of Indian companies also trade on the London and other European stock exchanges. As global financial markets become accustomed to Indian software companies, other Indian companies will find it easier to meet global accounting and financial standards and raise money in global

Table 10.1 Selected Indian ADRs* traded in the United States

Company	Symbol	Business/industry
Wipro	WIT	Information systems/software
Satyam Computers	SIFY	Information systems/software
Infosystems	INFY	Information systems/software
Silverline Technologies	SLT	Information systems/software
Rediff	REDF	Information systems/software
ICICI	IC	Financial institution
ICICI Bank	IBN	Financial institution
Dr Reddy's Laboratories	RDY	Pharmaceuticals

*American Depository Receipts.

markets. A great deal of this capital (hard currency) flows back to India, supporting the Indian rupee. A strong rupee means that the Reserve Bank of India can keep interest rates lower than without these export earnings. Such lower interest rates are likely to encourage growth throughout the economy; higher interest rates would put a damper on the economy.

Technology transfer from the software and other leading-edge industries will benefit some areas and industries earlier than others. India is a country with a large amount of geographical disparity, and there are certain areas that are doing much better than others in per capita income. In the north, the states near New Delhi are likely to benefit a great deal from the software industry and to see great growth over the next few years, especially in Punjab and Haryana. In the west, we have high growth in Gujarat and Maharashtra (areas near the Mumbai region). In the south, the city of Bangalore is considered the Silicon Valley of India. However, there is another city, Hyderabad in Andhra Pradesh, that is growing rapidly as a result of the high growth of the software industry. Indeed, there is likely to be higher growth throughout India, but certainly first in the major cities and their surrounding urban areas.

There are also likely to be industry variations in initial increases in growth rates. Driven by the software industry, the computer hardware industry will continue to see very high rates of growth. The growth of the software industry will also continue to add to the pressure on infrastructure. For example, in the telecommunications industry, India is rapidly being wired-up with large amounts of fibre-optic cable. Indeed, among developing countries the amount of fibre-optic cables being laid in India is second only to China (Jayaram, 2001). Consequently, many rural and most urban centres in India are expected to have broadband access via fibre-optic cables in the near future. The widespread availability of broadband will lead to a major increase in economic growth rates as these communities are connected to the global economy. There continues to be great shortages of electrical power in India, but many large businesses have already invested in decentralized electrical generating equipment. There will continue to be great demand for electrical power and for generating sets in India. In addition, logistics and transportation services will also have to grow very rapidly as a result. This includes much higher growth rates in air, rail and road travel and transport. Also, currently a large part of India's food production is wasted because of poor transportation and a poor distribution infrastructure.

The middle class in India is growing rapidly. It has been estimated that India now has somewhere between 70 and 90 million people living at the same standard as the US middle class.[3] The rapid growth in this class will mean a considerable jump in the demand for consumer goods. As there are considerable inefficiencies in the distribution and retail of consumer goods, there is a great potential in India for large retail chains. Another area which is likely to witness great growth is Banking and Finance, especially as inadequate banking and consumer finance is holding back much development in India. Box 10.1 summarizes these changes in Indian business due to economic deregulation and the application of the new internet-based technologies.

Box 10.1 Deregulation, technology and changes in the Indian economy

- Continuing deregulation and external opening of the Indian economy
- Deregulation promotes technology and technology encourages deregulation
- Application of new Internet-based technologies in India
- Rapid growth of the software and other selected high-tech industries
- Growth of India as a global centre for labour-intensive services
- Improvements in communications infrastructure in India
- Rise of an improved banking and financial system
- Emergence of the Indian middle class as a world-class consuming sector
- More efficient markets and reduced transactions costs
- Efficiency-improving restructuring and consolidation in many Indian industries

In summary, it is clear that India's economic role in the world is going to grow. Economic deregulation changes the structure of incentives and supporting social institutions in an economy and, thus, can lead to significant increases in the economic growth rate (e.g. Olson, 1996). India has great human and technological resources; it has strong technology in a number of areas, and it will continue to have 'islands of excellence' that will continue to spread. India's real long-term annual growth rate used to be only 3–4 per cent a few decades ago. Now it is more like 7 per cent reflecting a quantitative jump in the last decade. With the widespread deployment of new Internet-based technologies and world-class management practices, barring any major disruptions like armed conflict or war, it can be expected that in this decade there is likely to be another significant jump in India's economic growth rate.

Changes in corporate structure and strategy

Commerce and industry in India has been a mixture of heavily regulated small and large private businesses and large state-run businesses. The larger private Indian businesses have been characterized by large, widely diversified business groups. For example, of the largest 5446 public companies in India, 1821 are group-affiliated companies which are, on average, 4.37 times the size of non-group affiliated companies (Khanna and Yafeh, 2000).

This dominance of group-affiliated companies is quite understandable as the Indian commercial environment has been much less than perfect and the Indian markets for many corporate inputs have been quite inefficient (see Table 10.2). This has meant that Indian companies found it more efficient to internalize many of these markets; that is, not buy these inputs from external suppliers, but manage them as company resources. Consequently, many Indian business houses have been widely diversified and vertically integrated (e.g. Khanna and Palepu, 2000).

Table 10.2 Relative market efficiency for corporate inputs in the Indian economy

Market	United States	Japan	India
Finance	Equity-focused Wide disclosure External monitoring Market for corporate control	Bank-focused High debt ratios Group monitoring Limited market for corporate control	Underdeveloped Illiquid Weak monitoring Very weak market for corporate control
Labour	Business education Highly mobile General skills	General education Not mobile Firm-specific skills	Low education Low mobility Low skill levels
Products	Liability laws Enforced Efficient information Activist consumers	Liability laws Enforced Efficient information Few activists	Limited liability Enforcement Little information Few activists
Regulation	Low Predictable Not corrupt	Moderate Predictable Low corruption	High Unpredictable High corruption

Source: Based on information in Khanna and Palepu (2000) and author's analysis.

However, as the Indian economy is progressively deregulated and feels the impact of new information technology, external markets are becoming increasingly more efficient and the advantages of vertical integration and diversification will decline and eventually disappear (e.g. Rajan, Servaes and Zingales, 2000). These changes will force significant and sometimes sudden restructuring among the large and major conglomerates in India. Next we consider the forces driving these changes in corporate structure.

Transactions costs and corporate boundaries

A business firm can be considered to be a nexus of formal and informal contracts between various stakeholders – that is, customers, employees, suppliers, investors and other interested parties. These formal and informal contracts govern the many tasks performed by a firm. Some of these tasks are performed by units and entities within a firm, while others may be performed outside the firm.[4] The firm manages internal tasks through bureaucratic hierarchies or other internal mechanisms; external tasks are managed through outsourcing contracts and market processes. The number of functions and types of resources that the firm acquires, develops and manages internally versus those that it hires or acquires externally on an as-needed basis depends on which total acquisition costs are lower; total acquisition costs reflecting production costs and the costs and risks of the transactions necessary to acquire such resources (e.g. Coase, 1937; Williamson, 1995). Thus, to the extent that internal and external costs are similar, the boundaries of a firm are determined by the differences in external versus internal transactions costs.

Transactions costs generally consist of search, negotiation, contracting and enforcement costs of establishing a business relationship so that exchange of goods, services and compensation can take place. Contracts that eliminate all opportunistic behaviour are difficult or impossible to write because of bounded rationality, performance measurement difficulties, and information differences (asymmetry) between the parties to a contract. Thus, some transactions are best carried out in a market economy while others must be internalized and conducted within a firm. If external markets for these resources are inefficient and associated transactions costs are high enough, the firm will find it more profitable to internalize such markets by making such resources a part of the firm (e.g. Caves, 1980). Thus, *ceteris paribus*, the boundaries of a firm depend inversely on the efficiency of external markets – the greater the inefficiency of external markets, the wider are a firm's boundaries.

Transactions costs and market efficiency also depend on supporting social institutions that can provide the formal and informal rules governing the exchange process (North, 1990). Such institutions reduce transactions costs by reducing uncertainty and by providing a stable framework for forming ethical and other expectations regarding the behaviour of participants in a transaction and limiting incidents of opportunistic activity. Transactions costs are likely to be higher in an uncertain and changing environment. The nature and efficacy of social institutions, such as the legal system and the ethical framework in a society, can also play a critical role in reducing uncertainty or the costs of search, negotiation or contracting (e.g. LaPorta *et al.*, 1997). Similarly, strong systems of public disclosure of financial information, business monitoring, investor protection and corporate governance required by well-developed capital markets can also reduce transactions costs (e.g. Levine, 1997). As many of these institutional features that reduce transactions costs are more likely to be found in high-trust societies, transactions costs are likely to be lower in such societies (Fukuyama, 1995). In addition, poorly functioning markets for corporate control, managerial talent and technology can increase transactions costs. High-trust societies are likely to have lower transactions costs, and commerce in such countries is less likely to be dominated by large well-diversified and vertically-integrated business groups. Indeed, empirical evidence suggests that business groups serve as risk-sharing mechanisms when capital markets are underdeveloped but not when capital markets are well-developed (e.g. Aoki, 1988; Caves, 1989).

Markets in developing countries for managerial talent, technology, corporate control and other goods and services are generally less efficient than similar markets in developed countries. Further, institutional frameworks for reducing transactions costs and risks are also likely to be poorly developed. Consequently, firms in developing countries are likely to be more widely diversified and vertically integrated as compared to similar firms in developed countries. Of course, a number of factors other than underdevelopment may also be associated with wide firm boundaries. For example, business firm boundaries may also be wide in low-trust societies compared to high-trust societies, as transactions costs are generally higher in the former.

Changing transactions costs

With economic deregulation, many transactions costs are expected to decline. Most deregulation reduces the costs associated with managing the many business interactions with the bureaucracy, and the replacement of bureaucratic controls with market mechanisms generally means a move towards more efficient markets.[5]

In addition to the drop in transactions costs with economic deregulation, transactions costs in India and most other countries are also dropping due to the rapid application and implementation of the new Internet-based technologies in businesses. These new technologies can reduce transactions costs significantly as they certainly reduce the search and information costs involved. In many cases these technologies can also reduce negotiation, contracting and monitoring costs associated with business transactions. Thus, the application and implementation of these new Internet-based technologies has led to significant shrinkage of corporate boundaries in most countries. As one consequence, corporate outsourcing has exploded with almost no regard to physical distance. As discussed earlier, India is a large beneficiary of this trend towards outsourcing many corporate services.

Generally, as in many other countries, economic deregulation and the simultaneous application of new Internet-based technologies can be expected to greatly reduce transactions costs and improve market efficiencies in India. These changes in transactions costs and market efficiencies have very significant implications for the structure of Indian business.

Strategic implications for business groups

As in many developing countries, large business firms in India have generally been widely diversified and vertically integrated. However, the advantages of wide diversification and vertical integration are generally likely to disappear as markets become more efficient. In such cases, the disadvantages of a group structure become important; for example, such groups suffer from bounded rationality in managing diverse operations, sheltering of weak operations, and appropriation of minority shareholders' assets. Consequently, many business groups will have to restructure and become more focused.[6] Given the increased level of competition brought by economic deregulation, such groups no longer can or should stay in large numbers of unrelated businesses. However, such changes in business structure can be very challenging!

The Indian private sector has a long history – a history that includes many prior periods of considerable change. In addition to large numbers of small businesses, the Indian economy has also been characterized by the presence of many large business groups. Table 10.3 lists the 10 largest business groups over two 30-year periods ending in 1999. As the table shows, these largest business groups are becoming more important in

Table 10.3 Top ten business groups in the Indian economy

Rank	1939	CAGR1[a]	1969	CAGR2[a]	1999
1	Tata (62)[b]	7.3%	Tata (505)	12.8%	Tata (22345)
2	Martin Burn (16)	7.8%	Birla (456)	9.6%	Wipro (18439)
3	Bird (12)		Martin Burn (153)		Ambani (16060)
4	Andrew Yule (12)		Bangur (104)		HCL (9275)
5	Inchcape (11)		Thapar (99)		Ranbaxy (7970)
6	ED Sassoon (10)		Nagarmull (96)		Bajaj (7667)
7	ACC (Tata) (9)		Mafatlal (93)		Aditya Birla (7114)
8	Begg (6)		ACC-Tata (90)		Hero (3715)
9	Oriental T&E (6)		Walchand (81)		Satyam (3210)
10	Dalmia (6)		Shriram (74)		Punj (3173)
Average	(15)	8.5%	(175)	14.4%	(9897)

Note: [a] CAGR = compound annual growth rate (1=1939–69; 2=1969–99); calculated for the average and for the groups that survived in the top ten from one period to the next. [b] Figures in parenthesis indicate the asset base in 1939 and 1969 and market capitalization in 1999, on 31 March, measured in crores of rupees (1 crore = 10 million).

Source: Piramal (2001) and author's calculations.

the Indian economy as their average size grew at a faster rate than the Indian economy. Further, there were few commonalities in these lists for 1939, 1969 and 1999. Tata was the only one in all three lists, while Martin Burn was in the 1939 and 1969 lists and Birla was in the 1969 and 1999 lists. All others were new in each list (Piramal, 2001). Thus, the current period of business change is not new. However, the rate of business change in the next five to ten years is likely to accelerate and this period will be marked by much turbulence. There are many reasons for this.

Not all markets in a country are likely to become efficient at the same rate, and firms would have to monitor these changes and assess their impact carefully. In addition, such firms would have to redefine their core competencies carefully in view of the changed market efficiencies. However, this process of restructuring is unlikely to be entirely smooth. As in the past, many large business groups may not stay large or may even disappear.

Some firms are likely to undertake these strategic reassessments faster and more accurately than others. Such firms are then likely to drive many of the slower adjusting firms out of business. An appropriate bankruptcy procedure and a well-functioning legal system that can enforce property rights must be in place. However, failing firms, especially if they are large, are likely to create political and economic disruption.

Regulators and policy-makers would have to stand firm under such circumstances against calls to slow, stop or reverse the deregulatory process. Unfortunately, there is little experience in India with large-scale bankruptcies that are more common in capitalist countries like the USA.

Another major channel for business restructuring is a liquid market in corporate control in an environment of well-developed financial markets. Friendly and hostile mergers and acquisitions are an important tool for corporate restructuring.[7] Well-developed financial systems and capital markets need legal systems of clearly defined and enforceable property rights, an independent accounting profession and associated regulations to encourage high levels of public disclosure, and market regulations to prevent self-dealing and misuse of fiduciary responsibilities and encourage the fair and equal treatment of all shareholders (e.g. Aggarwal and Harper, 2001; Black, 2000).

In addition to enforceable and well-defined property rights, the development of financial and capital markets also depends on a reliable and transparent system of regulation. Such regulation is needed to ensure adequate and uniform disclosure, to prevent self-dealing and other misuses of fiduciary responsibilities in the financial industry, and to ensure fairness in financial markets and the financial soundness of securities exchanges and firms in the financial system. The orderly development of financial markets generally involves the progressive introduction of trading in securities of increasing risk and complexity. For example, the first securities markets generally involve short-term securities issued by the government, followed by trading in longer-term government debt instruments, bonds and shares of well-known and large private firms, bonds and shares of smaller lesser-known private firms, and then derivative securities. In addition to facilitating corporate restructuring, efficient financial markets have been shown to encourage economic growth (e.g. King and Levine, 1993; Levine, 1998; Rajan and Zingales, 1998).

In many cases, corporate restructuring in India is likely to be impeded by the lack of professional management, especially in family-run firms. While the situation is changing, a number of large business houses are still run by members of the controlling families. Lack of professional management might not only impede the analysis of strategic changes, it might also act as a deterrent to potential changes in corporate control. Economic deregulation and the development of financial markets are likely to hasten the transition from family control and management to non-owner professional management. Owners and corporate managers are likely to find it increasingly difficult to ignore the rights of minority shareholders.

Simultaneously, there is likely to be consolidation in many industries as the smaller and less-efficient players are likely to be driven out. It would be difficult for many of the smaller participants in a market to have the economies of scale necessary to deploy many of the new technologies, invest in building national brands, or develop the R&D capabilities necessary to survive in a deregulated business environment. Box 10.2 summarizes these transaction-cost-driven changes in the Indian business structure.

Box 10.2 Transactions costs and business structure in the indian economy

- A firm is a nexus of contracts for inputs – some external and some internal
- The nature of resource acquisitions depends on transactions costs:
  - costs of search, negotiation, contracting and enforcement
  - need to account for opportunism, uncertainty and incomplete contracts
  - information asymmetry, bounded rationality, measurement problems
  - transactions costs also depend on social norms and legal institutions
  - transactions costs depend on completeness and efficiency of markets
  - some transactions costs are lower in markets, others lower within firms
- Boundaries of the firm depend on transactions costs/market efficiency; on changes in technology, the social/legal framework and market efficiency; and on reduced transactions costs and more efficient markets
- Reduced scope and more focus among large businesses in India; increased bankruptcies and M&A in India
- Efficiency-improving restructuring and consolidation in many Indian industries

Discussion

The adoption of new technologies and advanced management know-how in India is being spearheaded by the development of the computer software and hardware industries. Technology and management practices from this lead industry leak out or are transferred to other industries in India making them more globally competitive. These developments further support and reinforce the process of economic deregulation. Thus, economic deregulation and the adoption of new technologies and superior management in India are mutually reinforcing forces that are likely to accelerate each other. This improvement in the institutional environment of business in India is equivalent to increases in organizational capital; increases that supplement reinvestments in the form of physical and financial capital. Thus, these changes can be expected to boost the rate of growth of the Indian economy.

However, these institutional changes also mean significant and often disruptive changes for large well-diversified and vertically-integrated business groups in India. As the markets for corporate inputs become more efficient, the advantages of diversification and vertical integration will decline and even disappear. With such changes, Indian businesses will have to become more focused and efficient. This transformation of large Indian businesses is unlikely to be entirely smooth and will require a continuing firm commitment to economic deregulation.

This analysis of the impact of new technologies and economic deregulation on the restructuring of large business houses and consolidation in many industries clearly has important implications for policy-makers and managers in business and industry. Each group faces many challenges and tasks to prepare for these upcoming changes. Given India's firm commitment to secular democracy and its vast differences in income and linguistic groupings, the process is likely to be hotly debated with many possibilities for opportunistic behaviour among the competing groups. It will require stable and responsible business and political leadership to ensure that India's economy continues to grow at the high rates of which it is capable.

Conclusions

This chapter has provided an assessment of the impact on Indian business of new technologies and economic deregulation in India. The rate of change facing Indian business is likely to accelerate as these new technologies and economic deregulation form a mutually reinforcing circle of forces. It has been shown that both the adoption of new technologies and economic deregulation will reduce transactions costs and make product and financial markets in India more efficient. This increase in market efficiency will then lead to a shrinking of the optimal size of corporate boundaries among Indian business firms. As most large business houses in India are currently widely diversified and in many cases vertically integrated, these changes will inevitably lead to a process of significant slimming down, focusing and restructuring of most of the major business houses. Simultaneously, there is likely to be consolidation in many industries.

However, these restructuring and consolidation processes are unlikely to be entirely smooth as many large business groups will be unable to survive in the new more competitive business environment. India has little experience with large-scale bankruptcies that are more common in countries like the USA. Failures of large firms are likely to be politically and economically disruptive, and India will need strong financial markets and bankruptcy laws and a robust commitment to the process of economic deregulation to survive the significant restructuring of its major businesses necessary for it to compete in global markets. Thus, unfortunately, it seems there are unlikely to be any gains in this area without some pain!

Notes

1 There is now much evidence that the institutional environment is a significant influence on national economic growth rates (e.g. Olson, 1996). Further, as evidenced by the great success of non-resident (overseas) Indians, the institutional environment rather than culture seems to be more important for Indian economic success (Pauly, 2000).

2 The elite Indian Institutes of Technology are now becoming well-known for graduating many CEOs of major US corporations and many of Silicon Valley's leading successes. See, for example, Ghosh (2001).

3 There are various estimates of the size of the Indian middle class. These estimates range from that of Ford of about 50 million that can afford an automobile, to over 200 million that can afford home appliances. The size estimate of the Indian middle class here seems close to a median number.

4 Some tasks may be performed jointly by units internal and external to a firm. However, this detailed issue is not critical to our analysis here.

5 However, there can be a temporary delay in achieving all of the benefits of economic deregulation and the application of new technology. In a deregulating environment like India, many social institutions that may reduce transactions costs may be in a state of flux. With deregulation, old institutional arrangements and frameworks become obsolete while it takes time to build stable new institutional arrangements and exchange frameworks. While economic deregulation and transition to a market economy is certain to reduce many transactions costs (such as those associated with the bureaucracy), such changes are also likely to temporarily exert an upward pressure on transactions costs.

6 This is starting to take place. See, for example, Joseph and Khandari (1999) on restructuring at Tata.

7 Domestic M&A is already surging in India (e.g. Khozem, 2001).

References

Aggarwal, R. (1999) 'Technology and Globalization as Mutually Reinforcing Forces', *Management International Review*, vol. 39(2), special issue, pp. 83–104.

Aggarwal, R. and Harper, J.T. (2001) 'Privatization and Business Valuation in Transition Economies', chapter 9 of L.L. Jacques and P.M. Valler (eds), *Financial Innovation and the Welfare of Nations*. Boston: Kluwer Academic, pp. 175–96.

Aoki, M. (1988) *Information, Incentives, and Bargaining in the Japanese Economy*. Cambridge: Cambridge University Press.

Black, B. (2000) 'The Core Institutions that Support Strong Securities Markets', *Business Lawyer*, vol. 55(4), August, pp. 1565–607.

Budhwar, P. (2001) 'Doing Business in India', *Thunderbird International Business Review*, vol. 43(4), July–August, pp. 549–68.

Caves, R.E. (1980) 'Industrial Organization, Corporate Strategy, and Structure: A Survey', *Journal of Economic Literature*, vol. 18(1), March, pp. 64–92.

Caves, R.E. (1989) 'International Differences in Industrial Organization', in R. Schmalensee and R. Willig (eds), *Handbook of Industrial Organization*. New York: North-Holland.

Coase, R.H. (1937) 'The Nature of the Firm', *Economica*, vol. 4(4), November, pp. 386–405.

Das, G. (2001) *India Unbound*. New York: Alfred A. Knopf.

Fukuyama, F. (1995) *Trust*. New York: The Free Press.

Ghosh, C. (2001) 'Boot Camp for Engineers', *Forbes*, 16 April, pp. 158–60.

Guha-Khasnobis, B. and Bhaduri, S.N. (2000) 'A Hallmark of India's New Economic Policy: Deregulation and Liberalization of the Financial Sector', *Journal of Asian Economies*, vol. 11(3), December, pp. 333–46.

Helpman, E. (1998) *General Purpose Technologies and Economic Growth*. Cambridge, Mass.: MIT Press.

Inglehart, R. (2000) 'Culture and Democracy', chapter 7 of L.E. Harrison and S.P. Huntington (eds), *Culture Matters: How Values Shape Human Progress*. New York: Basic Books, pp. 80–97.

Jayaram, A. (2001) 'Wiring Up', *Business World*, 26 March, pp. 50–7.

Jishnu, L., Jayaram, A. and Joseph, S. (1999) 'The World is Here: The WTO Clock is Ticking', *Business World*, 29 November, pp. 20–6.

Joseph, T. and Khandari, M. (1999) 'Remaking Tata', *Business World*, 13 September, pp. 17–25.

Joshi, V. and Little, I.M.D. (1996) *India's Economic Reforms 1991–2001*. Oxford: Clarendon Press.

Khanna, T. and Palepu, K. (2000) 'Is Group Affiliation Profitable in Emerging Markets? An Analysis of Diversified Indian Groups', *Journal of Finance*, vol. 55(2), April, pp. 867–91.

Khanna, T. and Yafeh, Y. (2000) 'Business Groups and Risk Sharing Around the World', mimeo, Harvard Business School (December).

King, R.G. and Levine, R. (1993) 'Finance and Growth: Schumpeter May be Right', *Quarterly Journal of Economics*, vol. 108(2), June, pp. 717–37.

LaPorta, R., Lopez-de-Silanes, F., Schlaifer A. and Vishny, R.W. (1997) 'Legal Determinants of External Finance', *Journal of Finance*, vol. 52(3), July, pp. 1131–55.

Levine, R. (1997) 'Financial Development and Economic Growth: Views and Agenda', *Journal of Economic Literature*, vol. 35(2), June, pp. 688–726.

Merchant, K. (2001) 'Domestic Companies Lead M&A Surge in India', *Financial Times*, 15 February, p. 12.
Mokyr, J. (1990) *The Lever of Riches*. New York: Oxford University Press.
North, D.C. (1990) *Institutions, Institutional Change, and Economic Performance*. Cambridge: Cambridge University Press.
Olson, M. (1996) 'Big Bills Left on the Sidewalk: Why Some Nations are Rich and Others Poor', *Journal of Economic Perspectives*, vol. 10(2), Spring, pp. 3–24.
Pauly, H. (2000) 'Passages', *Milwaukee Magazine*, May, pp. 58–65.
Percy, C. (1992) 'South Asia's Take off', *Foreign Affairs*, vol. 71(5), December, pp. 166–74.
Piramal, G. (2001) 'The History Lesson', *Business World*, 8 January, p. 62.
Porter, M.E. (1990) *The Competitive Advantage of Nations*. New York: The Free Press.
Rajan, R.G. and Zingales, L. (1998) 'Financial Dependence and Growth', *American Economic Review*, vol. 88(3), June, pp. 559–86.
Rajan, R.G., Servaes, H. and Zingales, L. (2000) 'The Cost of Diversity: The Diversification Discount and Inefficient Investments', *Journal of Finance*, vol. 55(1), March, pp. 35–80.
Williamson, O.E. (1985) *The Economic Institutions of Capitalism*. New York: Free Press.

11

Intangibles, Technology Trade and India: Challenges in a Knowledge-Based Economy

*Anupam Srivastava and Seema Gahlaut**

Introduction

Globalization has brought about profound changes in national and multinational perspectives regarding the regulation of technology transfers and exports. The integration of markets and ideas, and the digital and electronic revolutions, has enabled transnational exchange of goods, services and information through intangible networks. Governments are now faced with new challenges for regulating these networks, whether for patent security or national security, or for controlling the proliferation of weapons of mass destruction (WMDs). Many of these challenges were underlined in a grim fashion on 11 September 2001, with the ease of terrorist planning and execution of attacks on key US targets including the 'Twin Towers' of the World Trade Centre in New York and the Pentagon in Washington, DC.

India has carved a growing niche for itself in the global IT and space markets. It also has significant capabilities in civilian and military spheres of nuclear, missile and chemical technologies. Although India has remained outside multilateral export control regimes, it has so far maintained an excellent record of controlling exports of WMD materials and technology.[1] However, as the country seeks to integrate itself with global markets and to allow greater private sector participation in the development of dual-use technologies, it will need to establish stronger and clearer regulations regarding technologies in general and

* Part of this research was made possible under a multi-year grant from the Carnegie Corporation of New York and another grant from the Ploughshares Fund.

intangible technology transfers in particular. This chapter will build upon the current perspectives and policies of the technological leaders – the United States and Europe – to identify the current perspectives and policies of the Indian government and industry. It will also provide specific recommendations for India with reference to this emerging area of international concern.

The changing definition of technology

'Information Revolution' is best understood as a result of a set of related technical innovations, together yielding substantial enhancements or even revolutionary change. Every two years the speed of silicon processors doubles and storage costs decline by half; every five years compression efficiency is squeezing data to one-thirtieth its size; and laser diode speeds, critical for fibre optics, are doubling every three and a half years.[2]

Esoteric technologies that previously appeared to have few civil applications are now being rapidly applied across a wide range of civilian activities, both for commercial and domestic purposes. These include the new digital technologies in everyday appliances, the application of GPS-based navigation systems to personal use, adoption of advanced telecommunications systems on a global basis, and extremely advanced sensors to support real-time production and control systems. At the same time, it has increased the vulnerability of information systems, whether commercial/civilian or military.

These new capabilities raise significant legal, economic and social questions about intellectual property rights, data security and integrity, privacy, as well as national security. For instance, technology-generation and innovation now involves teams of scientists and engineers working across national boundaries. Much of such innovation takes place in the private sector, and within companies that are multinational not just in terms of their location but also in their workforce, management, ownership and target markets.[3] Consequently, perhaps the most disquieting effect on national and international organizations and regulations is that this information/digital revolution has made the traditional distinctions and boundaries fuzzy. This includes the traditional distinctions between domestic and foreign firms, civilian and military applications, and national and commercial security considerations.

Governments, firms and countries that are technology-generators realize that the various forms of a controlled technology can be transferred to unauthorized entities through *intangible means* (Internet, telephone, fax, e-mail, and even scientific conferences and seminars). Further, since the end of the Cold War, entities interested in acquiring weapons of mass destruction have shifted their emphasis to obtaining technology and knowledge ('know-how' and 'know-why') rather than materials; that is, *acquisition of intangible technology*.

The current definition of 'technology', therefore, includes specific information required for the development, production or use of a product. This includes both *technical data* (blueprints, plans, diagrams, engineering and design specification, written or recorded manuals and instructions etc.) and *technical assistance* (training, skills, instruction, consulting services etc.), recorded as CDs, ROMs and computer disks.

American and European perspectives and policies

Controls on technology have existed since ancient times, but have had similar justifications throughout – denial to specific domestic and/or foreign entities in order to punish them or to curtail their capabilities.

Throughout the Cold War, for instance, an elaborate set of controls were devised under an agreement called the COCOM (Coordinating Committee for Multilateral Export Controls), to restrict the export of advanced information technology, especially computer technology, to the Soviet bloc and other non-allies. The strategy was intended to preserve a US/allied qualitative edge.[4] It worked well because support for this research by hi-tech firms came from the governments, and governments were also the sole buyers. Over time, this agreement was complemented with the establishment of a series of informal suppliers' cartels dealing with nuclear, chemical and missile technologies.[5] At the end of the Cold War, the Communist threat ended. However, the technology-control regime found a renewed focus in the proliferation of the Weapons of Mass Destruction (WMDs) – especially as it related to the acquisition of dangerous or sensitive technologies by 'countries of concern', most of whom happened to be located in the developing world.

Now, with the Information Revolution, new realities have emerged. First, the vital distinction between the United States and its allies versus the rest of the world is no longer in the area of computing power, but in the technology to make that power matter:

> The key factors supporting US superiority in information technology have shifted from the computing capacity necessary for higher-level operations to the tacit knowledge and system integration information

necessary to convert highly advanced computer operations into productive tools for civilian and military industries.[6]

Second, product innovation and development is now mostly in the commercial sector, where competitiveness drives constant upgrades and newer applications for an expanding, near-global market. Third, governments and militaries are being forced to cut costs and buy the more efficient and more innovative commercial-off-the-shelf (COTS) technologies. They can no longer maintain unilateral control over information and technologies. Fourth, militaries are no longer the sole or even the largest buyers for most high-technology companies.[7] The research, development and export activities of the technology companies are multinational by necessity, and national security communities are finding it hard to ensure that foreign countries and entities do not derive unauthorized benefits from this situational dynamic.

These trends are visible in most countries that are harnessing the knowledge economy for enhancing or maintaining their technological lead. Therefore, governments around the world are working with industry to devise means of controlling the flow of information and technology to entities (firms, states and terrorists) that they view as rivals or threats, either in commercial or in military terms.

Governments in the developed world are beginning to formulate laws and procedures to deal with this problem in two ways:

1 regulating the transfer of controlled technology by intangible means (i.e., *how* is the controlled technology being transferred); and
2 regulating the transfer and export of intangible technology (i.e., defining *what* is intangible technology).

The first concern leads directly into the realm traditionally defined under national secrecy laws that govern foreign travel, public presentations and private communications and interaction of persons involved in projects that are national security-related. However, as the preceding section has indicated, such projects now involve persons from the industry as well as academia, and may include nationals of several countries. Hence the scope of such secrecy laws is being stretched and challenged. The second concern is a direct result of the boom in dual-use technologies, that are apparently harmless in themselves yet have the capability to contribute to military uses in certain configurations. Advanced technology, thus, is becoming increasingly dual-use in nature, and *fungible* (or changeable) both *between* as well as *across* the civilian and military sectors. In this context, globalization is inducing technology diffusion and making it possible for technologically and economically backward states, firms and terrorist-groups to acquire the know-how and know-why that undergird civilian products and to configure them for military uses. The following discussion provides a concise overview of US efforts to legislate and implement Intangible Technology Transfers (ITT), along with analogous efforts in key states (UK, Germany, Russia).

In the United States, trade in conventional weapons is governed by the Arms Export Control Act, and includes export and import of defense articles and services, including data transfers. The policies and procedures are based on the International Traffic in Arms Regulations (ITAR). The US Munitions List, contained in ITAR, enumerates the specific controlled items. Trade in dual-use commodities, technologies and software is controlled under the Commerce Control List (CCL) of the Export Administration Regulations (EAR). Regulations related to technology controls are also contained in the Atomic Energy Act, the Foreign Assistance Act, and numerous other Executive Orders and sanctions laws. US laws do not distinguish between transfers of technology by hardcopy or by electronic means such as the Internet.

Above and beyond these controls on technology, the US Immigration and Naturalization Service (INS) has the authority to exclude foreign nationals, who wish to enter the United States from engaging in any activity that would violate or evade US export control laws. Common understanding of such activity includes study at US academic institutions, teaching, conducting research, participating in exchange programmes, and attending meetings and conferences, being temporarily trained or employed. Consular offices work with a Technology Alert List (TAL) to screen out visa applicants from countries listed in the 'projects, regions, and countries of concern' list of the EAR.

Once the foreign national is inside the United States, tracking his activities becomes difficult, and the onus is on academic institutions or businesses to inform the authorities. Thus, colleges and universities are required to inform the authorities if the foreign national changes his/her academic programme. A foreign national's access to sensitive technology is generally referred to as a 'deemed export'. Academic institutions, government agencies and firms are required to apply for technical data licences if a foreign national will have such access during his/her research, site-visit, training or employment. The United States also controls all electronic transmissions of non-public data that may be accessed by a foreign national. Thus any upload to public electronic bulletin boards, Internet file transfer protocol, or World Wide Web sites is considered actionable as an export. EAR also controls such 'exports' of encryption software.

Security regulations at US government laboratories have been tightened since the Wen ho Lee case of 2000 at the Los Alamos National Laboratory in New Mexico in the United States. Under the Department of Energy (DOE) Action Plan, its scientists are now required to get a special license to visit foreign states, have their presentations reviewed for secret information, and report their interactions with foreign individuals including e-mail and telephone correspondence. Access of personnel to computers with sensitive information is now more tightly controlled, as is access of foreigners to DOE facilities. Counter-intelligence regulations now include polygraphs, financial reviews and rigorous background investigations for persons working in 'special access programmes'. Major European allies of the United States, e.g. UK, Germany, France and Russia, have progressively modified their respective export control regulations to incorporate a similar (broader) definition of technology, as well as policies that control intangible technology transfers.[8]

Indian policies on control of dual-use technology transfers[9]

As India seeks to integrate itself with global markets and allows greater private sector participation in the development of dual-use technologies, it will need to establish stronger and clearer regulations regarding intangible technologies. The idea is to foster a creative and constructive analysis of current rules and institutions, both national and international, that regulate international trade in information or knowledge-based products and services. The following analysis sketches out the framework of Indian regulations regarding exports and transfers of dual-use technologies.

The broad guidelines for the non-proliferation-related export control policy in India is governed primarily by the following laws:[10]

(a) The Foreign Trade (Development and Regulation) Act 1992 (no. 22 of 1992), or the FTDR.[11]
(b) The Atomic Energy Act 1962 (no. 33 of 1962)[12]
(c) The Customs Act 1962

Taken together, these three Acts govern the formulation of the export–import (EXIM) policy. The most recent is the EXIM policy for the period 2002–7, notified on 31 March 2002. Detailed explication of the policy and procedures governing controls on imports and exports is given in the following four documents:[13]

1 Export and Import Policy, 1 April 2002–31 March 2007
2 Handbook of Procedures Vol. I
3 Handbook of Procedures Vol. II (Standard Input–Output and Value addition Norms)
4 ITC (HS) Classification of Export and Import Items (includes item-wise import and export policy based on ITC (HS) classification)

End-user certification is required for both imports and exports. This is the responsibility of the importer or exporter, and inconsistencies in this statement render the person liable for punishment, including a suspension of his export license as well as his Importer – Exporter Code (IEC) number.

For nuclear-related items, the Department of Atomic Energy (DAE) is developing a database to keep track of activities of exporters and importers of nuclear materials. This task is relatively easy now because of the very small number of actors (government and private) involved in the nuclear sector currently, but is expected to grow in the coming years as India gears up to allow more exports and imports of nuclear materials and technologies. If charged under the Atomic Energy Act of 1962, espionage and treason charges might be brought against a person charged with selling or exporting proscribed information and materials without proper certification and authorization. According to one DAE official, the actual scope of this Act is quite 'draconian'.

With reference to exports of chemical-biological materials, the government is moving ahead on implementing its CWC obligations as well as establishing procedures for stricter pre-shipment verification of end-use by the chemical and pharmaceutical industries.[14] The Directorate General of Foreign Trade (DGFT) mandates that exporters/industry representatives personally visit the importing site and assure themselves of the *bona fides* of the concerned entity, *before* applying for the export license.[15] This is in addition to the signed undertaking with end-use/end-user declaration that is required with license applications, which includes certification that the end-user 'shall not cause the items, or replicas, or derivatives thereof, to be re-transferred/sold without the knowledge and consent of the Government of India.'

The penalties for violation of laws through misleading information and/or attempts to import or export proscribed goods, include provisions under the Code of Criminal Procedure (CrPC.) 1973. These may range from confiscation of goods, monetary fines, to imprisonment. Civil laws (Indian Penal Code or IPC.) apply in suspension of license under the Conservation of Foreign Exchange and Prevention of Smuggling Activities (COFEPOSA) Act 1974.

On the issue of controls on cryptographic technology, so far the cryptography situation in India largely remains in the developmental stage.[16] However, cryptographic products compatible with DOS and Windows environments and customized cryptographic products have been designed and produced for commercial use. The issue of public and private key usage is under the purview of the Joint Cipher Bureau, the agency responsible for handling all encryption-related technology in India. The Department of Telecommunications does not permit encrypted signals to be transmitted over their network, but this does not appear to be enforced. Further, Indian industry largely produces PC software products containing cryptography at the 56-bit level. Stronger systems are not yet produced in India.

In 1998, the Bharatiya Janata Party (BJP)-led government introduced the Indian Information Technology Act. This legislation requires all internet service providers (ISPs) to monitor all traffic passing through their servers, making traffic, including the plain text of encrypted traffic, available to 'properly constituted authorities' for 'valid reasons of security'. Properly constituted authorities include the Central Bureau of Investigation (CBI), the Intelligence Bureau (IB) and the Research and Analysis Wing (RAW).

Indian perspectives for the future

At least four trends make it imperative that India takes up the issue of intangible technology regulations seriously. First, it has carved a growing niche for itself in the global IT and space markets. It also has significant capabilities in civilian and military spheres of nuclear, missile, and chemical technologies. It is soon to become a significant exporter of civilian technologies and manpower resources that will have dual-use capabilities. Second, the growing government–industry–academia partnership in charting out avenues for technology export and import will also mean greater privatization of technologies that were formerly controlled exclusively by the government-owned labs and public sector enterprises.

Third, years of denials by technology control regimes forced indigenization of technologies, allowing scientific and technological institutions to develop new products and processes in the areas of nuclear energy and medicine, lasers, computers, composite materials and so forth. Government laboratories are now waking up to the need of patenting these to facilitate both farming out to domestic firms for production, or for exports through international collaborations. Finally, understanding and responding proactively to foreign regulations, that is, the deemed exports laws, will be needed to ensure that their negative effect on technology exports to India can be minimized. We elaborate on each of these issues below.

Growing exports

The Centre for Development of Advanced Computing, a scientific society of GOI's Department of Electronics, will export its fastest supercomputer *Param 10000* to Russia and Singapore, while various units of its predecessor, *Param 8600*, had been sold to educational entities in Germany, Canada and more.[17] India has also offered its expertise in information technology and research and development sectors to ASEAN nations.[18] The offer is to train 100 ASEAN nominees in 'state-of-the-art IT applications' to signal its arrival as an increasingly important actor at the frontiers of the new global economy in the software and IT fields. Competitiveness in costs,[19] and strong attention to quality, is positioning Indian IT firms to enhance and maintain a strong international presence in at least the near to medium term. The Indian Space Research Organization (ISRO) has initiated similar programmes in recent years to impart technical data as well as training to a number of countries in Southeast Asia and Latin America.

It is clear that exports, particularly knowledge-intensive exports, will comprise an important component of India's growth in the period ahead. Further, in the budgets for fiscal years 2000 and 2001, the Finance Minister has increased the cap on foreign equity investment in, and ownership of, firms in the Indian private sector. Similarly, many restrictions on Indian firms buying up foreign firms or situating their subsidiary units abroad have been relaxed substantially.

All of the above indicate the growing scope and use of data transfers and sharing of restricted information amongst the many subsidiary units of a company located across more than one country. The need for security of commercial information transfers thus becomes evident. In turn, it is also likely to raise the demand from the industry for higher caps on permissible encryption technology employed to secure data transfers. GOI will thus have to not only formulate a robust mechanism governing such capabilities, but also the avenue to permit regular industry input and interaction such that this mechanism remains effective without stifling commercial efficiency.

Government–industry–academia partnership

Economic liberalization in the 1990s has not only removed the 'licence-raj' in India, it has also removed the mental block in the government that hindered a two-way consultation and coordinating dialogue between government and industry. Various industry associations as well as domestic and foreign experts are now consulted regularly before and after framing laws related to development, exports and imports of technology. The Confederation of Indian Industry (CII) and the Federation of Indian Chambers of Commerce and Industry (FICCI) now routinely offer recommendations on new initiatives and regulations, and offer partnership to key ministries and agencies.

Technology-intensive exports accounted for about 10 to 14 per cent of the country's total exports in fiscal year 2000, but the export of real technology accounted for much less – around 0.15 per cent. A recent Government of India (GOI)-sponsored experts' panel came to the conclusion that the performance could be enhanced and India should be able to achieve a 2 per cent share of global technology exports without much problem.[20] Therefore, the CII joined hands with the Department of Scientific and Industrial Research (DSIR), in the Science and Technology Ministry, to float a new organization to exclusively promote the export of technologies and technology-intensive products. The new Technology Export Development Organization targets a wide spectrum of activities ranging from computers, software, telecommunications, micro-electronics, civil aviation, biotechnology, enviro-tech to agro-tech, chemicals and allied fields, and plant designing, detailed engineering, turnkey projects and manpower training.[21]

In recent years, the CII has signed memoranda of understanding and agreements with several institutions and agencies, both abroad and within the country, to promote technology, technology development and technology networking in industry. The CII has also suggested to the GOI that a single agency could be created to regulate the biotech sector.[22]

Meanwhile, in March 2000, the Federation of Indian Chambers of Commerce and Industry (FICCI) and the United States India Business Council (USIBC) have signed a protocol, known as the 'knowledge-trade initiative'. The initiative seeks to encourage, support and facilitate a business-to-business dialogue with a view to accelerating the synergy between the two countries in the areas of knowledge-based products and services. It will centre on four substantive areas – e-commerce regulation and taxation; trade and services; intellectual property; and movement of natural persons.[23] It intends to explore if the existing rules and institutions are adequate enough to support the kind of trade in intangible products and ideas that are visualized in a dynamic bilateral business relationship, while taking into account the overall global economic development.

A similar government–industry partnership has been proposed and promoted in the field of defense production and in defence-related high-technology R&D. For instance, under the Defence Electronics and Research Laboratory project called *Samyukta*, the software for *Agni, Prithvi, Akash* and *Trishul* missiles was developed by a team from a private company. DERL and the company not only signed a 10-year after-sales support contract, but also a non-disclosure agreement binding each contracting party.[24]

Patenting technologies developed in government laboratories

The CII has suggested to the GOI that there is a growing need to institute a strong IPR regime with modern patent offices and training programmes to develop human resources for manning these offices.[25] The government laboratories are also taking this issue seriously. BARC and DRDO have recently begun to patent technologies they have developed over the years, especially in the spheres of composite materials and processing.

According to recent reports, Indian scientists and engineers working on the light combat aircraft (LCA) had to innovate and invent processes and products that could overcome the technology controls imposed by the United States and other advanced industrial countries. It was only in November 2000 that the Director of the National Aerospace Laboratories formulated the idea of culling this tacit knowledge spread across several defence/aerospace laboratories, to register these as patents.[26] Similarly, anticipating problems with the continued supply of enriched uranium, BARC has developed MOX technology – a mixed oxide fuel comprising natural uranium and plutonium dioxide – a major step forward in nuclear fuel self-sufficiency, which makes BARC an equal partner with Russia in the development of mixed thorium-uranium dioxide fuel. Moreover, India has also been successful with the Fast Breeder Test Reactor in using a thorium blanket along with MOX.[27] These technologies are being patented and will contribute to the design of small power reactors that India hopes to export in the near future.

At the same time, Indian technology-generators are becoming more aware of the dangers of using foreign technology in domestic programmes. In January 1999, the Defense Research and Development Organization (DRDO) issued an 'alert' against all network security software developed in the United States because it is concerned that US export law permits only such software to be exported that can be broken by the US National Security Agency:

> When various multinational companies go around peddling 'secure communication software' products to gullible Indian customers, they conveniently neglect to mention this aspect of the US export law.[28]

Deemed exports

India has targeted $50 billion in software and services exports by fiscal year 2005. Regardless of how much of this target will be realized and on time, a large portion is likely to be IT manpower exports. This export will be either directly to industries in the developed world, or indirectly as Indian graduate students in foreign universities specialize in hi-tech research and move on to jobs in those areas. Government–industry partnership in technology R&D in the developed world increasingly involves academic institutions, and a growing majority of researchers are foreigners. Currently, regulations regarding deemed exports (transfer of intangible technology to foreigners) assume that the foreigner will 'take away' existing technology through training/research. However, increasingly, foreigners not only provide the labour, but also produce/invent new technologies. Ownership and patent over such technologies is of growing interest to manpower exporters such as India.

On the other side, as foreign companies set up subsidiaries in India, the transfer of technology between the parent company and the subsidiary firm is also regulated under deemed export provisions. If India is considered a 'country of concern', technology imports by subsidiaries may be severely curtailed in dual-use technologies. Further, a related concern of foreign governments is securing legitimate/authorized uses and reexports of their technology in and from India. If Indian regulations are found wanting, and not considered adequate for protection of technologies related to national security or patent security, technology imports are likely to be hampered.

However, currently available information indicates that Indian policy-makers are less concerned about framing regulations that define 'deemed export' of Indian technology. They are more concerned about the potentially negative impact the 'deemed export' legislation of other countries can have on Indian manpower exports. A related concern is that unclear foreign laws might put Indian inventors/innovators working abroad at a disadvantage.

Future challenges for a knowledge-based Indian economy

Globally, regulations on intangible technology transfers are converging around similar definitions and procedures within multilateral security regimes, involving both (a) internal security measures within government and private institutions involved in technology-generation, and (b) measures to ensure authorized end-uses of controlled technology.

Although India has remained outside of multilateral export control regimes, so far it has maintained an excellent record of controlling exports of WMD materials and technology. Emerging trends indicate positive but complex changes in India's technology profile. However, continuing to remain outside the multilateral security regimes might put the country at a disadvantage both as a technology importer and as a technology exporter. The following recommendations are provided with a view to bridging the gap between the Indian stance and the median viewpoint within multilateral security regimes.

First, the next-generation IT industry, that is, moving from data entry and software support to systems development and integration, will require a more proactive regulatory approach towards intangibles exports. As all technology-generators before it, India will have to clarify and/or frame policies that safeguard the interests of domestic innovators while ensuring that government controls on technology exports to forbidden entities and destinations do not get undermined by private interests.

Second, as Indian companies have begun to acquire foreign firms, the GOI will have to ensure that information exchange between parent and subsidiary firms does not bypass export control regulations. In this context, it would be worthwhile to reexamine the possibility that the ultimate responsibility – to ensure that there be no illegal exports to countries of concern – should be seen as residing with the Indian exporters.[29]

Third, security checks for scientists and persons involved in projects of national importance are currently based on the antiquated Official Secrecy Act. Authorizations for foreign travel and clearance for presentations are done routinely, giving dedicated scientists benefit of the doubt. However, as the liberalized economy provides greater opportunities for private sector jobs and incentives for making profits through industrial espionage, more than routine checks might be needed for constant and impartial vigilance.

Challenges to ITT in liberal economies

Experiences of advanced industrial states (and Russia) suggest that it is difficult to control tacit knowledge. Russia is clearly faced with the problem of containing the problem of brain-drain *vis-à-vis* its unemployed or underpaid nuclear and missile scientists. Similarly, the globalization of knowledge, for instance through the education of foreign-born students in the United States, appears to be eroding the tacit knowledge superiority of that country.

Current legislation in most advanced countries allows exemption from deemed-export licensing for fundamental research, based on the notions of academic freedom and freedom of speech,[30] but many national security experts are concerned that this is a loophole that should be plugged.[31] However, scientific knowledge requires interaction and collaborations to grow:

> Before US satellite export controls were tightened about two years ago, the US satellite industry won about 70 percent of all international contracts for communications satellites. Today, US export controls have proven so onerous that foreign customers avoid them entirely by eschewing US manufacturers and eliminating the use of US components.[32]

An associated problem is that of restricting access of foreign students:

> In the United States, 52 percent of masters and doctoral students in technical and engineering sciences are foreigners, mainly from India and China. Should these students be prohibited from taking classes? In Russia, foreign students are a major source of funding for poor universities.[33]

As India establishes centres of excellence in information technology and biotechnology and promotes industry–university partnerships, it too will attract a growing number of researchers from around the world who might pose similar challenges for distinguishing between academic and defense-related research.

Beyond that is the issue of government controls over communications of its citizens, which raises both ethical (privacy) and logistical (implementation) problems. For instance, currently US aerospace companies are paying for US Department of Defense personnel to monitor their conversations with their clients.[34] Can firms in developing countries like India implement such costly practices? Already the Indian provisions on encryption and IT, that is, mandating that ISPs provide access to intelligence and national security agencies, is under fire.

In conclusion, the GOI will need to allocate an adequate national policy priority, and the requisite technical and financial resources, to bear upon its regulatory framework regarding intangible technology transfers. A useful benchmark to consider in this process would be the standards adopted by the advanced industrial democracies of Western Europe. At a minimum, establishing such a regulatory policy framework, and transparency and information-sharing regarding the operations within this framework, would facilitate greater technological collaboration and industrial cooperation with Western Europe as well as the United States. This might also facilitate greater investment of technology-imbedded capital as well as commercial collaboration from the advanced industrial countries. Until such time as India remains outside of the various multilateral export control agreements and arrangements, such a regulatory framework might also facilitate negotiations for pragmatic bilateral solutions that provide for greater cooperation with India.

Contemporary globalization, and the increasing dual-use applications of advanced technology, makes it clear that 'supply-side' controls over intangible transfers alone are unlikely to suffice. It becomes equally imperative to take into account the 'demand-side' of the equation, including the legitimate end uses of such capabilities by the relevant actors within the international community. It also underscores the urgent need for international policy coordination in devising the regulatory mechanisms, as well as criteria for monitoring, verification and enforcement regarding intangible technology transfers.

For a developing economy such as India, with an emphasis on knowledge-intensive exports, this provides an avenue to adopt a proactive approach. This would enable the country to join the international mainstream of such discussions, and provide it an opportunity to shape the agenda of the discussions from within rather than risk becoming a marginalized voice of dissent from without. Finally, the Indian experience would likely be of particular relevance to developing countries as they each determine the optimum balance, or the 'dynamic equilibrium', between allocating precious national technical and financial resources to intangible technology controls, and pursuing other legitimate developmental imperatives.

Notes

1 For a description of Indian policy on dual-use export controls, see Richard T. Cupitt and Seema Gahlaut, 'Indian and US Perspectives on Nonproliferation Export Controls', in Gary Bertsch, Seema Gahlaut and Anupam Srivastava (eds), *Engaging India: US Strategic Relations with the World's Largest Democracy* (New York: Routledge, 1999).

2 Jeffrey R. Cooper, 'Information Processing, Command, Control, and Communications: Impacts of the Information Revolution', paper presented at the 14th Annual Ottawa NACD Verification Symposium, 4 March 1997.

3 For a fuller discussion on this issue in the United States, see *Final Report of the Defense Science Board Task Force on Globalization and Security*, Office of the Under Secretary of Defense for Acquisition and Technology, December 1999.

4 For a US perspective on Cold War-era export controls, see John J. Hamre, 'Establishing an Effective Modern Framework for Export Controls', Hearing before the Committee on Banking, Housing and Urban Affairs, United States Senate, 14 February 2001. For a European perspective, see 'On the way to a new export control system', a project report prepared by the Danish Agency for Trade and Industry, September 2000 (ISBN 87–90704–92–4).

5 These were the Nuclear Suppliers' Group (NSG) in 1974, the Australia Group (AG) in 1985, and the Missile Technology Control Regime (MTCR) in 1987. COCOM was disbanded in 1994, and a new agreement called the Wassenaar Arrangement (WA) was established in 1996. It seeks to control dual-use technologies that are not controlled by the other three.

6 For details on this theme, see 'A New Approach to National Security and Information Technology', Final Report of the Centre for Strategic and International Studies, March 2001.

7 For details on this theme, see Anupam Srivastava, 'Technology Flows and U.S. Leadership', in *The Monitor: Trade, Technology, and Security in the 21st Century*, vol. 6, no. 2, Spring 2000, pp. 33–5.

8 For more information on how the UK, Germany, Russia and the United States view the problem of controlling intangible exports, see *The Monitor: Nonproliferation, Demilitarization and Arms Control*, Summer 2000, available at www.uga.edu/cits

9 This section borrows from Seema Gahlaut, 'Indian Policy on Controlling Exports of Sensitive Technologies: Achievements and Future Challenges', in Kenneth A. Cutshaw (ed.), *Corporate Counsel's Guide to Doing Business in India* (Chesterland, OH: Business Laws, Inc., 2001).

10 Other laws that govern the movement of persons and things in and out of India include the Explosive Substances Act 1908; the Narcotic Drugs and Psychotropic Substances Act 1985; and the Environment Protection Act 1986.

11 This replaced the Imports and Exports (Control) Act 1947.

12 The Atomic Energy Act 1962 replaced the Atomic Energy Act of 1948. The latest amendment to the 1962 Act is Notification AEA/27/1/95-ER dated March 1995.

13 All four are publications of the Ministry of Commerce, Government of India.

14 The government of India (GOI) signed the Chemical Weapons Convention in 1993, and became a full participant in the activities of the Organization for the Prohibition of Chemical Weapons (OPCW) at The Hague in Netherlands since 1999.

15 One interviewee admitted to the author that his company was forced to conduct recent site visits to China and Saudi Arabia under this requirement.

16 Report written by the Information Technology Group at the Department of Electronics, Government of India.

17 'Param 10000 for export', George Iype, *Rediff* on the Net, 28 September 1999.

18 'India offers IT expertise to ASEAN countries', *Rediff*, 29 July 2000.

19 See, for instance, Nayan Chanda, 'Software Exports: How Precarious is India's Perch?', *Far Eastern Economic Review*, 12 April 2001. See also, Gary H. Anthes and Jaikumar Vijayan, 'Lessons from India, Inc.', 2 April 2001 <http://www.computerworld.com/cwi/story/0,1199,NAV47 STO59083,00.html>

20 'CII, DSIR to float organization for technology export', *The Hindu*, 1 April 2000.

21 See, for instance, M. R. Srinivasan, 'An S&T agenda', *The Hindu*, 24 June 2000.
22 'CII drafts blueprint to give reforms a boost', *Rediff* (Money), 27 March 2001.
23 'The Knowledge Trade Initiative: Indo-US Protocol on Knowledge-Based Products and Services', Reported in *The Hindu*, 24 March 2000.
24 B.G. Prakash, 'The Millennium and Satellite as Weapon', *The Indian Defense Review*, Oct.–Dec. 2000.
25 'CII drafts blueprint to give reforms a boost', *Rediff* (Money), 27 March 2001.
26 B.G. Prakash, 'Dreams Lighten in LCA: NAL Discovers Knowledge Management', *The Indian Defense Review*, Oct.–Dec. 2000.
27 D N Moorty, 'India perfects MOX fuel N-tech', *Indian Express*, 14 October 2000.
28 Quoted at http://www.ciphers.de/crypto/legal.html reg_by_country
29 This is known as the catch-all provision in most countries, and assumes that the onus of gathering information on end-users lies with the exporter, above and beyond the prescribed minimum required by domestic law.
30 For discussions on how the United States, the United Kingdom and Germany define 'fundamental research', see the articles on intangible technology transfers by Timothy Williams, Bridget Butt and Andreas Klein (respectively) in the Summer 2000 issue of *The Monitor: Nonproliferation, Demilitarization and Arms Control*, available at www.uga.edu/cits
31 See for instance the report, 'Interagency Review of the Export Licensing Process for Foreign National Visitors', US Department of State, April 2000.
32 Neal Lane, 'The Openness Imperative', *Foreign Policy*, March–April 2001.
33 'Globalization and Control of Intangible Technology Transfers: A Major Challenge to Export Controls in the 21st Century', CITS Working Paper, *The Monitor: Nonproliferation, Demilitarization and Arms Control*, Summer 2000, available at www.uga.edu/cits
34 Colin Clark, 'Report Calls for Clearer US Knowledge Export Rules', *Space News*, 10 April 2000.

12

Industrial Relations and Increasing Globalization: A Case Study of India*

C.S. Venkata Ratnam

Introduction

This chapter briefly traces some of the key issues and trends in industrial relations in the wake of globalization, with particular reference to India. Specifically it deals briefly with (1) the historical development of industrial relations and the corresponding legal framework in the country, (2) examines the impact of liberalization on industrial relations in the context of increasing globalization, (3) discusses employment, employment security and social safety nets, (4) key issues concerning multinationals, privatization and export processing zones, (5) new approaches to work organization, (6) international labour standards, (7) recent trends and developments in collective bargaining, workers' participation, labour laws and industrial relations, and (8) finally discusses the role of trade unions, employers organizations and the government in developing systems of industrial relations.

Historical overview of industrial relations in the country

India was predominantly an agricultural economy until Independence in 1947. Even after Independence, the First Five-Year Plan (1951–56) laid emphasis on agriculture. With the Second Five-Year Plan (1956–61) there was a shift towards heavy industrialization. The share of agriculture in gross domestic product (GDP) declined from about 56 per cent in 1950–51 to less than 30 per cent in 1990–99, while the share of industry rose from

* This material was originally prepared for the International Confederation of Free Trade Unions, Singapore, for a conference organized in Manila, August 2000.

15.6 per cent to 24.7 per cent, and of the services sector from 29 per cent to 45 per cent during the corresponding period. It produces a wide variety of industrial and consumer goods. Over the years, the dependence of the performance of the industrial sector on that of the agriculture sector has reduced. India also did not feel the crunch of recession or the wrath of business cycles. Its middle class has grown to over 250 million, a vast enough market, even as over 220 million continue to live below the poverty line.

The industrialization strategies and industrial policies followed in India from Independence to the mid-1980s aimed at development and a faster growth rate, but emphasized regulation rather than development. To protect the domestic industry from foreign competition and promote self-reliance, import substitution was encouraged as a state policy. But import substitution and export promotion were not viewed as two sides of the same coin. Instead, the plethora of controls led to the erection of barriers of entry and exit, to pygmies being treated as giants, and monopolies denying economies of scale to Indian enterprises, to tariff and other restrictions on imports and exports, distortions in pricing of inputs and outputs, and to a sheltered market and protective regime based on political patronage. The end result was the creation of preemptive capacities to minimize or avert competition, gross underutilization of capacities, profiteering by trading in licences and quotas (for state-controlled raw material, etc.) and to collusion rather than competition. Thus, too much government in business and industry, and vice versa, eventually proved detrimental and counterproductive to the conduct of state affairs as also to industry and commerce, workers and unions and consumers and the community.

The following could be considered important developments in the historical evolution of industrial relations in the country. The listing is illustrative, not exhaustive:

- 1910s: Conditions of workers deteriorate in the wake of the First World War.
- 1920s: Indian Trade Unions Act and Workmen's' Compensation Act were enacted.
- 1930s: Royal Commission on Labour expressed concern over the conditions of labour. 1931 Karachi Session of Indian National Congress passed resolution about the need to safeguard the interests of workers through legislation and other means
- 1940s: First Tripartite Indian Labour Conference under colonial rule. Political independence. Key legislations – Industrial Employment

(Standing Orders) Act; Bill for regularization of Dockworkers' employment conditions; Industrial Disputes Act; Minimum Wages Act; Payment of Wages Act, etc.

- 1950s: India became a republic with 'justice, liberty and equality' as the ideals enshrined in the Constitution. Economic Planning started. Emphasis on heavy industrialization, which laid the foundation for the growth of public sector to attain the commanding heights of the economy. Nationalization of Imperial Bank and insurance companies. Spirit of voluntarism resulted in the signing of several declarations and codes of practice.
- 1960s: Spirit of voluntarism continued. The first National Commission was constituted in 1996. It submitted its report in 1969 and recommend revamp of labour policy, law and institutions.
- 1970s: Rise of regional parties resulted in further splits in trade unions and increase in tension between the Centre and the States on industrial-relations issues. Series of nationalizations (banks, coking and non-coking mines, foreign oil companies, etc.). International emergency and abridgement, temporarily, of workers rights and suppression of all India strikes in transport sector. Industrial strife and militancy in West Bengal. Amendments to Industrial Disputes Act to secure job protection. Amendments to Constitution defining duties of citizen and making workers participation in management as part of directive principles of state policy. Tripartism atrophied.
- 1980s: Longest strike in textile industry. Militant phase of union movement in Western India. Piecemeal amendments to labour legislations.
- 1990s: Economic liberalization. Legislature, Judiciary and Executive soft towards unorganized sector and strict towards organized sector. Introduction of family pension and a series of other welfare facilities in favour of disadvantaged and vulnerable groups. Encouragement to private sector, foreign direct investment and multinational companies. Setting up of the Second National Commission with a mandate to recommend changes in labour law taking into account the changes in product market characteristics and an umbrella legislation for the unorganized sector. Seeds sown to liberalize, privatize and globalize the economy. Some unions are very critical about the new policies being anti-labour aimed at deindustrialization and privatization of profits and nationalization of losses. National Renewal Fund (NRF) was established as a social safety net for workers affected by restructuring, but it served mostly to 'voluntarily' retire the 'surplus' staff. Employees' Pension Fund was set up to provide pension benefits to employees and

their family from out of provident fund contributions. Rejuvenation of tripartite dialogue: frequent interactions, but hardly concrete actions. Banking and financial sector reforms and liberalization of banking and insurance. Dismantling of public monopolies in utilities and setting up of regulatory agencies (in electricity, telecom besides stock markets and insurance) began.

- 2000: Modern Bakeries became the first central public enterprise to have been privatized. Significant privatization and second-generation structural adjustment reforms planned. Interest rate on provident fund was reduced by 1 per cent.

Development of a legal framework

India inherited the colonial legacy in its legal framework governing industrial relations:

1. The legal framework for industrial relations has focused more narrowly on the 8 per cent workforce in the organized sector. The plethora laws provided little coverage and extended protection to a microscopic minority of the workforce. There is a need to make the coverage universal and get over the mindset of linking applicability of laws based on number of persons employed and amount of wage/salary drawn. In 2000, Maharashtra Government took the initiative to provide, for the first time in the country, protection for domestic workers. Now they are entitled to 15 days paid leave in a year and travel concessions for home leave, and a bonus on the eve of Diwali, the festival of lights.
2. In the pre-liberalization era, labour was protected in the labour market and capital in the capital market. In the post-liberalization era, both labour and capital feel less protected or unprotected. Indian employers feel that after globalization instead of realizing the opportunities in the global markets, they are not able to compete even in domestic markets. In the past, consumer interests were neglected. Now, when consumers' rights clash with those of labour and capital, the rights of consumers and the community seem to prevail over those of labour and capital.
3. Indian legislation has focused on dispute resolution rather than on the maintenance of sound labour–management relations. The stress was on adjudication rather than on collective bargaining.
4. Labour laws in India are considered not friendly to change and prevent easy exit. Workers and unions are concerned that easy exit means unbridled freedom to hire and fire. Therefore, even sick and unviable

units – whatever be the reasons, and labour may not be the reason in most cases – are forced to carry on, at least with a view to protecting the jobs and livelihood of those employed.

5 In practice, most labour laws have neither found favour with employers nor with workers and their unions. Both employers and unions agree on the need for changing labour laws, but not on the content of the proposed changes.

Impact of liberalization on industrial relations

There are doubts whether liberalization leads to competition. The spate of mergers and acquisitions and the domination of a few brands or major players once again points out that in the era of globalization economies of scale matter. Though the number of players has increased, first due to easing of entry barriers, eventually few remained. Actually in sectors like oil, air transport, telecom, banking and so on there are doubts whether the number of players who will ultimately remain in the field will be fewer than before liberalization. Domestic enterprises find it hard to compete even in home territory because they were not used to competition and mostly operated in a seller's market. Both captains of industry and academics are asking the government to liberalize the domestic economy first before globalization. The private sector is asking to privatize the private sector first. The Eleventh Finance Commission (2000) observed that economic liberalization did not result in liberalization of the public sector enterprises (PSEs) from government controls. The dismantling of public monopolies is accompanied by the creation of regulatory authorities which are subordinate to political and bureaucratic control. When ineffective regulatory authorities replace inefficient public utilities a bad situation could become worse. A chamber executive remarked, 'The evil of licensing is replaced by the devil of tendering.'

Technology is redefining competition. After nationalization, nationalized banks expanded their branch networks into remote and rural areas and onto virtually every street corner in urban areas. Eventually many of them have become unviable. The new, foreign banks with the latest technology are open 24 hours a day, seven days a week, with automatic teller machines. Furthermore, e-banking has metamorphosed the banking industry. The CEO of a foreign bank said that, 'we cannot have a bank on every street corner, but we can place one in every home, every phone, every personal computer.'

The government has drafted competition policy, but what is important is that the politicians, bureaucrats and the captains of industry should walk the talk and talk the walk. They, thus far, usually walk the walk and talk the talk. The basic ingredients of a competitive environment are still missing in India: demand–supply mismatch, information imperfections and skewed distribution of resources, including purchasing power, and so on. More importantly the much-needed reforms have eluded the public sector in the real sense. With most of the inputs and infrastructure – social and physical – still being in the hands of the government, unless the public sector gears up, the rest of the economy cannot. Still critical are reforms in the three pivotal areas: legislature, judicial and executive.

How different countries respond to the requirements of competitiveness may be a function of the stage of development and state of health of the economy concerned. While different countries have different critical moments in history and pursue divergent industrialization strategies, it is usually observed that

> a less developed country begins the process of industrialization by creating some initial conditions conducive to investment. In industrial relations terms, it may translate to low wages and low unionization. This situation may attract initial investments by firms, which can take advantage of such labour market conditions. However, with increased investment, the initial labour market conditions inevitably change and there are pressures for higher wages and possibly unionization ... This reduces the initial advantage which attracted the new investment in the first place so other inducements are needed in order to retain attraction to external investors. This creates a critical juncture in the development process and places pressure on the existing industrial relations system.

The State at this stage may introduce measures such as wage controls and union repression or, alternatively, take recourse to upgrading skills and linking wages to productivity and shift to high-wage, high-value industrialization where workers get higher wages but wage costs become competitive. Wage controls cannot be prolonged and union suppression by itself cannot guarantee economic success. Therefore wage controls and union repression is uncalled for. Still, it becomes necessary to create conditions necessary for inducing investment. Putting jobs before growth is like putting the cart before the horse. There can be growth without jobs, but there cannot be any new jobs without growth.

The policies of the international financial institutions (IFIs) have been generally criticized by most trade unions as being responsible for exacerbating the existing economic and social problems and worsening the industrial relations climate. Their policies would not have mattered much if the country learnt to live within its means. IFI policies have affected urban India directly; rural India has felt an indirect and secondary impact. Some sectors were affected badly and others not. Overall, India's performance in the post-reform period was below its potential economically, and negative on the social dimension. The key criticism against IFIs came from Jalan (1996): (a) the policies are imposed without recognizing the diversity of country circumstances; (b) the 'one size fits all' approach could cause more harm than good; (c) the social costs of adjustment and reform could cross the limits of tolerance; and (d) the 'success' stories are few as evident from the need for repetitive IMF assistance in most countries.

However, India did not face economic crisis because of the following reasons: (a) it retained its controls on capital outflows, (b) it did not go in for full convertibility, (c) foreign exchange remittances by migrant workers of Indian origin and investments by NRIs has contributed to the accumulation of a sizeable exchange reserve base, and, (d) it has moved cautiously and on a case-by-case basis, so far, in matters concerning dis-investments, etc.

Industrialization strategies affect the social and labour aspects and influence human resource and industrial relations policies at both the macro and micro levels. These strategies and policies should thus contend with the contextual elements of the labour market to ensure that economic and industrial development is in harmony with social and human progress. From this perspective it is necessary to take stock of some key factors of the labour market in India:

1 The supply of labour is in excess of demand. The demand for capital is in excess of its supply. Open unemployment may be less for reasons explained later, but underemployment and disguised unemployment are high, and so is the incidence of poverty and illiteracy. Taken together, these conditions make it difficult, regardless of the nature of laws, strength of trade unions and the political will of the government, to ensure job security and social protection to workers.
2 Economic growth *per se* had not been low in India. But the modest 'Hindu' rate of growth was nullified by the increase in the rate of growth of India's population. Hard economic conditions make even the government, with the best intentions, less able to safeguard the interests of workers and unions.
3 Gupta's (1999) comparative study of the periods 1983–91 and 1991–98 points out that given the same rate of growth of GDP during both

periods, the incidence of poverty increased and the rate of growth in employment decelerated:

Changes in poverty and employment and GDP

Years	Changes in poverty ratio	Employment growth rate	GDP growth rate
1983–91	−3.1	1.6	5.6
1991–98	+2.7	1.1	5.7

4 The famous economist Alfred Marshall observed that if either or both the input and output of labour can be substituted, and if the proportionate costs of labour are high, unions' capacity to achieve positive outcomes at the bargaining table would be low.
5 Political instability may be perceived by some as good, at least in the short term, for workers and unions because in such circumstances governments would not take hard decisions and become increasingly sensitive to political risks.
6 Pressure to restructure public finances is leading the government to disinvest in public enterprises mainly with a view to achieving targets set for revenue realization and to reduce revenue deficits. Selling capital assets to wipe out revenue deficits is not a wise proposition.

Employment and employment security

India has a workforce of 380 million. Over 92 per cent are in the unorganized sector, 54 per cent self-employed and 15 per cent in regular wage employment. 70 per cent of the 8 per cent employed in the organized sector (over 19 million out of about 28 million) are employed by the government. Between 1991–98 organized-sector employment grow by 1.6 million, from 26.7 million to 28.3 million. The private sector, which employs less than one-third of the organized-sector labour, accounted for two-thirds of the growth in employment during the 1990s.

Although job creation has been a major concern and objective of successive five-year plans, over the years fewer jobs have been created per Rs1 million investment. In the 1950s a leading economist, Mahalanobis, estimated that an investment of Rs10 million in infrastructure would create 11000 jobs. During 1980–87, 70 000 jobs were added in the organized sector and capital formation in the public and private sectors exceeded Rs700 000 million. Thus, Rs10 million investment generated just one job on average in the organized sector through the greater part of the 1980s and the 1990s. The Nhava Sheva port in the public sector employs 10 times less manpower, and its wages cost as a proportion of total operating expenditure was 12 times less than that of Mumbai. A public sector integrated steel mill set up at Visakhapatnam generated less than 14 000 jobs on an investment of over Rs80 000 million. A new joint sector steel mill proposed to be set up in Madhya Pradesh soon will entail an investment of Rs400 000 million with jobs for 400 persons. Two successive changes in technology in telecom – manual mechanical to electrical mechanical to digital cellular – reduced manpower intensity by a factor of 500, that is, one person is needed where 500 worked previously.

Labour legislation during the Emergency in 1975 required prior permission from the government for lay-offs, retrenchment and closures. Such legal provisions, court judgments and past collective agreements between employers and trade unions created rigidities in the labour market. However, higher adjustment costs seem to have reduced the demand for labour in firms in the organized sector, irrespective of whether they are owned by the government or by the private sector. A study of employment trends in 34 Indian industries, according to the Annual Survey of Industries Data for the period 1976–82, pointed to a long-term decline in the demand for labour at around 17.5 per cent (Fallon and Lucas, 1991). The study noted significant inter-industry variations: employment reduced by more than 5 per cent in 25 of the 35 industries, and by more than 15 per cent in seven of them. The rate of decline in employment was estimated to be over 33 per cent in textiles. Another study of 34 firms in Mumbai estimated an average reduction in employment of 20.5 per cent during the 1980s (Workers' Solidarity Centre, 1989).

Several studies (Goyal, 1984; Ramaswamy, 1988; Fransen, 1991) have pointed to a shift in employment from the organized to the unorganized sector through subcontracting, and the emergence of a typical employment practices where those who work for the organization do not have an employment relationship but a contractual relationship (Mathur, 1989), during the 1980s. The process was accentuated in the 1990s and beyond.

Most enterprises in India have not invested in R&D. In the wake of economic liberalization their prime concern had been to become competitive through cost and/or price-based strategies rather than through value addition, creativity and innovation. This means, in the traditional manufacturing sector, producing four times more output with one-fourth of the workforce. In the process, even if wages are doubled, wage bills would be halved and wage costs would be reduced by eight times. During the 1990s, Sandoz reduced its workforce by 90 per cent. BOC reduced its workforce to one-fourth and multiplied its business four times. The private sector steel giant, Tata Steel is planning to reduce its 70 000 workforce to one-third and its public sector counterpart, Steel Authority of India Limited, is planning to reduce its 220 000 workforce by one-third. The approximately 240 central public-sector undertakings divested over 300 000 persons through voluntary retirement between 1999–2000. The Vth Central Pay Commission (India, 1997) recommended a reduction of civil service personnel by 30 per cent. The Eleventh Finance Commission (2000) recommended 'right-sizing' of the manpower in civil service and observed that several state governments had emphasized the need to reduce the size of the government, privatize state parastatals, avoid filling vacancies arising out of retirement and recommended across-the-board cuts of 10 per cent in non-salary and non-plan expenditure.

Against this background, the country was desperately searching for a middle way. As the then Prime Minister P.V. Narasimha Rao observed while addressing the World Economic Forum at Davos, Switzerland, in February 1994:

> Change has to be accepted as a result of deliberate and objective thinking. At the same time, those who wear the shoe and know where it pinches should have full say in deciding how to mend it. In the new-found enthusiasm for change, governments should not go overboard and plunge large chunks of their people into mass misery; they have no right to do that . . . each society has to find its own 'middle way' suited to its genius and circumstances; and, this should be the approach that accepts change.

Clearly, the Indian case – in terms of both the magnitude and complexity of the challenges – calls for a different strategy than elsewhere. Given its strong democratic traditions and vibrant – even if weak and splintered – trade union movement, neither an autocratic nor a corporatist approach to industrial relations will work. Striking a balance between the requirements of economic development and social progress has not been an easy task. Therefore, the search for the 'middle way' has not yet produced a model that could be operationalized on a wider scale, though in certain specific local contexts at the firm/plant level both unions and managements have been able to work out an equilibrium and get into agreements on workforce reductions with higher compensations than envisaged in law. Courts have upheld that if the choice was between retrenchment and closure, agreements proposing the former could not be considered unlawful.

The female participation rate in employment continues to be low: in the 1991 census, the usual status of male and female work participation rates (WPR) were 51.9 and 18.4 per cent respectively. There are large variations in female work participation between states: higher in southern and western states and lower in northern (except Himachal Pradesh) and eastern states. The information technology industry has displaced women from some of the traditional office jobs and created new jobs in relatively skilled areas. The higher in hierarchy, the greater the gender gap in employment. However, women's presence is improving in higher education, particularly professional courses. While a few states have started implementing a 30 per cent reservation for women in education and employment, in many others there is just lip sympathy. In the information technology industry, there is a move to change laws like banning women in night work, and so forth, so that they can also access jobs like telework, which is usually round the clock.

Overall, there is a growing concern about jobless growth, reduction of employment and employment opportunities in the formal sector, and growth of actualization and normalization of labour. Employers argue for labour-law reform to increase flexibility, but unions apprehend that flexibility may be a euphemism for freedom for employers to hire and fire the workforce at will. Unions argue that 92 per cent of the workforce in the unorganized sector do not enjoy job security or income protection and afford employers full flexibility. Why can't they leave the remaining 8 per cent as it is?

There has been no change in the legal framework concerning employment security during post-liberalization, although both the executive and the judiciary have been showing greater responsiveness to product market requirements. In the past, even if companies were sick, employees' interests were secure. Now, even financially sound companies' jobs are not as secure. The dominant message is that enterprise success is a necessary but not sufficient condition for job security. Employers argue about the need for proactive restructuring and justify downsizing as right-sizing to prevent sickness and protect the interests of the surviving workforce in the organized sector. Given the absence of a social safety net for the poor and the unemployed, the incidence of poverty is found to be higher among the employed than the unemployed. The really poor do not have the capacity to withhold their labour and settle for any job, even if at below minimum wage sub-subsistence levels.

Even though the rate of population growth has marginally receded, given its age distribution the labour force in the country is projected to increase faster than the demand for labour, over the next decade. Trade unions have a tough time dealing with situations/arguments such as the following:

- Restructuring leads to job loss, but non-restructuring does not avoid job loss. Job losses could be more with non-restructuring than with restructuring.
- New technologies shed few jobs and create many more jobs. For example, two successive revolutions in telecom – manual mechanical to electrical mechanical to digital cellular – have not only resulted in a loss of a few thousand group C and group D jobs in some public sector enterprises and erstwhile departmental undertakings, but also created a few million jobs in STD/ISD booths throughout the country's five lakh of villages. Quality is an issue, but for many who do not have a source of livelihood, some job is better than no job.
- In some units – for example Sindri Fertilizers, some units of Indian Drugs and Pharmaceuticals Limited and Hindustan Engineering Corporation – for years there has been either no or low production. In others – for instance Hyderabad Allwyn before privatization and Kolar Gold Mines – the losses would have been less without production than with production. The expenses in Kolar Gold Mines to produce 10 gms of gold is nearly five times the market price of the same quantity of gold!
- Both consumers and the community do not seem to be interested in how a company is addressing itself to social issues like job-creation; they are more interested in the cost and quality of products and services. Before the private interests of the consumers and the public interests of community the sectional interests of both employers/managements and workers/unions seem to have become secondary and/or relegated to the background. Consumer courts are awarding judgments undermining the rights of workers, and public interest litigation on matters concerning environmental aspects and so forth is resulting in verdicts from courts that subordinate the interests of the workers and employers to those of the wider public and the community.

Social safety nets

Social safety nets are needed to provide assistance for maintenance of self and family livelihoods:

- in the event of loss or reduction of income due to involuntary unemployment;
- for sickness, accident, invalidity or old age; and for
- redistribution of cash, goods and social services – education, health, subsidized housing, etc. – to the needier sections of the community.

In discussing social safety nets in the context of increasing globalization, we are mainly concerned with protecting the unemployed, particularly those rendered redundant due to restructuring to make the units that once employed them viable and competitive. Workforce reduction should be the last, not the first, option in restructuring. The provision of a social safety net is intended to take care of the vulnerable. Its purpose is to soften the adverse effects on people displaced rather than make it easy for employers to get rid of people. It is underlined that not all sick units need to be closed or resort to workforce reductions. Similarly, profitable companies may not remain viable forever unless they adapt to changes. Productivity can be improved not only by denominator management (producing more with less people) or numerator management (producing more with same people) where the demand for the product is not a problem. Experience reveals that the absence of a social safety net has resulted in delay or deferment of restructuring Timely restructuring will avoid closure of units which would otherwise become chronically sick and lead to closure. The longer the delay in restructuring, the more the problem becomes chronic with a harsher impact on labour.

In today's environment a large number of companies want to reduce the numbers they employ and cut the proportionate cost of their wage bills. Capital wants security for investment, and labour is concerned about the security of their jobs, career and earnings. Income security is being considered an alternative to job security. The record of the National Renewal Fund created in 1991 is very disappointing, to say the least. It spent over 90 per cent of the money to 'voluntarily' retire over 200 000 employees in the PSEs. Its record in retraining and redeployment is also very poor, and it has not done anything with regard to area regeneration and the setting up of an insurance fund (see, ILO, 1999).

Several countries believed in the recent past that redundancy could be fought with retraining and redeployment, that skills obsolescence could be prevented or minimized through retraining. But, when the new jobs created are less than the job losses plus the fresh additions to the labour force, redeployment will solve the problem of the redundant workers, not the unemployment of the fresh entrants into the job market. Several companies have introduced schemes which provide for higher compensation than mandated by law to those who they want to retire 'voluntarily'. Profitable companies proactively restructuring their enterprises through downsizing may have the capacity and commitment to develop such schemes and make payments over and above the legally mandated amounts. Sick and closed units, however, do not often have the capacity and commitment to meet even the legal dues. Social safety nets are needed to take care of the latter kind of situations. In addition, social safety nets should also be considered for providing subsidies to economically depressed industries facing financial difficulties with a view to arresting the resort to retrenchment of workers (to keep employment within enterprises during the crisis period) or to minimize the effects of lay-offs and retrenchments. The subsidies could cover in certain cases up to 50 per cent of the total wages costs. It is also suggested that the relevant laws could be amended such that in the unfortunate event of closure of enterprises and liquidation procedures, workers' wages, severance pay, unpaid bonuses, and other such outstanding payments to workers should be the first charge on the liquidated asset of the concerned company or group of enterprises (ICFTU-APRO, 1998).

Key issues: MNCs, privatization and EPZs

Multinational companies (MNCs)

India needs foreign capital and technology resources to make optimal use of its vast natural and human resources and the growing market for a variety of products and services. In the past, policies of self-reliance and import substitution have restricted the scope for foreign capital and technology in India. The deregulation of foreign investment in India in 1991 can be considered a watershed. The extent of the liberalization of the foreign investment climate is such that as a chief executive of one multinational corporation (MNC) put it, 'India has opened up its economy with far less discrimination than most other countries, including the Fast East and Southeast Asian economies at their commensurate state of development.'

The role of foreign direct investment (FDI) was recognized in India in its Industrial Policy Resolutions of 1948 and 1956, but not in key industries reserved for exclusive growth in the public sector till recently. The Foreign Exchange Regulation Act (FERA) 1973 put, barring certain exceptions, a ceiling of 40 per cent on foreign equity participation in India. Multinational corporations which did not want to dilute their stake were asked to leave the country. FERA has been a major deterrent to FDI in India till the New Industrial Policy (NIP).

The Monopolies and Restrictive Trade Practices Act was amended to remove the threshold limits with respect to assets in monopolies and dominant undertakings. In high-priority industries, automatic permission is given for foreign equity upto 51 per cent as against 40 per cent in the past, and in several areas foreign equity is permitted up to 100 per cent. Prior approval of the government is not required for investment in deregulated industries, which gives greater flexibility for MNCs in planning and diversification of their operations. Specific high technology and priority industries are now given automatic approval to conclude foreign technology agreements within certain guidelines. Permission is no longer necessary for the hiring of foreign technicians and the testing of locally developed technologies outside India. The Foreign Investment Promotion Board (FIPB) negotiates with a number of large international firms to promote substantial investment, improved access to advanced technology and world markets. The bulk of the investment in the 1991 liberalization era came through the FIPB route.

The MNCs main attraction in India is the size of its market, which is not only large but is also growing phenomenally. A vibrant democracy, credible judicial system, strong bureaucracy, competitive human and natural resources and continued popularity of English as a business language add to its attraction. However, Indian industry is concerned about the role of MNCs in India. In the early 1990s many Indian companies had gone in for foreign tie-ups and many foreign companies increased their stake in Indian operations. A spate of mergers and acquisitions followed, and many alliances became soured too soon for one reason or another. Established Indian brand names began to be replaced by foreign brand names, and a new practice of charging royalties for use of brand names, both by Indian and foreign companies, created a flutter.

Both Indian employers and Indian trade unions share broadly similar perceptions and concerns about MNCs, which are along the following lines:

1 MNCs are interested in capturing the Indian market, but not building a manufacturing base in India. Continued import of components is cited as proof of this allegation. Very few Indian companies are being developed to source parts from India for the overseas operations of MNCs. The few exceptions include the case of Sundaram Fasterners which won a global tender to supply radiator caps for General Motors, worldwide. Recently, the Korean auto giant, Daewoo, also acquired control over its Indian operations, increasing its stake to develop the Indian unit to source auto components for its worldwide operations. It failed, though, due to a financial crisis at headquarters.

2 MNCs focus on the short term rather the long term. MNCs are keen to generate and repatriate profits quickly.
3 MNCs are using India as a dumping ground to bring in technology and products which are being phased out in their home countries.
4 MNCs use Indian partners to get a foothold in India on a 50/50 or 40/40 basis, to get sanctions and approval quickly and, once established, they seek to edge out, as in the camel and the Arab story, the Indian partner.
5 MNCs have joint ventures with Indian partners but simultaneously have also set up competitive areas without Indian partnership. They thus use the insights obtained through partnership with Indian companies in joint ventures to secure unfair advantage and allow the 100 per cent subsidiary to compete with the joint venture firm.
6 For significant proportion of MNCs, investment is going either to increase stakes in existing businesses or mainly to supply second-hand machinery to relocate their obsolete plant in the home/third country in India. Such MNC investment does not generate many new jobs, and even if they are created, obsolete plants become the breeding grounds for sickness and therefore job losses.
7 MNCs arrive in India like 'cowboys'. They hastily choose a partner, make mistakes and want to break the relationship. Alternatively, they get into alliances with different Indian companies for different product lines. In the telecom sector, all the joint venture partners who bid for cellular tenders have parted company already; and alliances in the deregulated airlines industry also soured.
8 MNCs cause deindustrialization. The ghost of the East India Company still haunts the average Indian psyche. The East India Company came for business and colonized the country for over two centuries and contributed to India's deindustrialization. Now in several industries even established Indian brand names are not able to face or withhold foreign competition due to the financial muscle and brand image of the foreign competitors in auto (two, three and four wheelers), electronics (television sets), white goods (refrigerators, for instance) and soft drinks.
9 An impression is gaining ground, in the wake of renegotiation of certain power projects, that MNCs have a tendency to pitch their investment costs and prices to a rather higher than necessary level, pushing prices onto the consumer. Confessions about huge development costs have bred suspicions about corruption. The Secretary-General of a Chamber of Commerce even told a public gathering: 'The evil of licensing of the past is now replaced by the devil of tendering.'

10 Recent revelations about foreign exchange irregularities involving ITC–BAT, and the controversy concerning control of ITC by its UK partner BAT has not augured well to boost the image of MNCs in India.

11 MNCs are capitalizing on the weaknesses of traditional and small-scale Indian businesses in rural areas by quickly patenting herbal products (plant based) and indigenous snacks (Bikaneer bhujia, for instance), spreading a feeling of distrust and hatred towards multinationals. The entry of some multinationals to businesses like salt (salt satyagraha launched by Mahatma Gandhi in Gujarat was an emotive movement during the freedom struggle) in border areas of Gujarat, agriculture (Cargill) in Karnataka, ecologically unfriendly chemicals projects in Goa (Dupont's pact with Thapers) and using India as a dumping ground for disposal of dangerous chemical wastes have rendered MNCs rather unwelcome in the minds of several Indians.

MNCs in Indian manufacturing, at least in the 1980s, paid wages higher than their local counterparts. This can be explained in terms of their tendency to employ qualitatively superior personnel than their local counterparts. Recently, when General Motors decided to set up its manufacturing facility in Vadodara, the Vadodara Industrial Employers' Union reportedly met its chief executive and pleaded with him that he should not upset the local wage structure in the interest of maintaining employer unity in the region which predominantly had many medium and small sector units paying modest wages.

Working conditions in MNCs are usually better than their counterparts in the private sector. A veteran trade union leader commented recently, 'trade unions hate multinationals, but like the pay and perks they offer' (Tulpule, 1993). In many large MNCs in the chemical and pharmaceuticals industry union density rates are higher than the national average. But in the electronics industry owned and controlled by the MNCs unionism is virtually absent, apart from a few private (ICIM, for instance) and several public sector electronics units characterized by fairly high union density rates.

Unionism is very strong in both the public and the private sector banks in India, but particularly in the foreign banks. Several foreign banks have registered unions, but some foreign bank managements do not recognize them. While some like the Grindleys Bank sign agreements with their unionized staff, and a few others have chosen to become members of the Indian Banks Association and thereby become party to the industry-level collective agreement, many foreign banks refuse to sign agreements directly with their unions. A proper measure of industrial relations in such organizations is not the number of mandays lost, but the number of court cases concerning unfair dismissals and other forms of discrimination and exploitation of labour in which they are embroiled.

Collective bargaining in India is size-specific. Large companies employing more than 300 persons typically have collective bargaining, except in insurance businesses. Till now MNCs do not operate in the insurance sector, though in future they may. Most MNCs, other than in electronics and a few in banking, have collective bargaining.

Japanese investors are apprehensive about the 'somehow' management of unions and industrial relations in the context of multiplicity, plurality, ambiguity and uncertainty. The Japanese MNC representatives of the Kansai Productivity Centre, who recently visited India, told the author that they would like to identify and deal with a collective bargaining agent direct. But, unfortunately, the union structures and the recognition policies do not provide such clear-cut arrangements for determining collective bargaining agents.

In India, generally higher wages and higher exploitation go together. In many cases, workers could secure higher wages relative to other firms in the region through hard bargaining and militant unionism. The same can be considered generally true of MNCs, particularly in chemicals and pharmaceuticals, but not so much in engineering industries and certainly not for MNCs in electronics. One MNC located on the outskirts of Delhi signed an agreement in 1995 with its individual employees to the effect that it would increase their emoluments by a few hundred rupees more if they agreed not to join/form a trade union.

Typical Indian private sector industrialists are comfortable with the enormous diversity in the legal framework of India and many of them somehow manage the complexity and uncertainty if they are clear about their purpose. MNCs, in contrast, generally expect clear-cut legal measures because they do not want to be seen as good corporate citizens at least with regard to legal compliance. MNCs provide the social security measures mandated by law – including provident fund and gratuity, but not pensions so far. Several MNCs, however, have pension funds for their senior executives, but not for workers.

The aim of the Tripartite Declaration of Principles on Social Policy and Code of Code for Multinational enterprises

> is to encourage positive contribution which MNCs can make to economic and social progress and to minimize and resolve the difficulties

> to which their various operations may give rise, taking into account the United Nations Resolutions advocating the Establishment of a New International Economic Order.
>
> (ILO, 1993, p. 8)

Workers' organizations are generally not consulted on investment matters, particularly in regard to policy changes on investment, assets and other operations of MNCs. The Indian National Trade Union Congress (INTUC) acknowledges that MNCs are expected to play an important role in promoting economic and social welfare, by providing more employment opportunities and raising the living standards of workers. The INTUC states that the establishment of MNCs is regulated in accordance with the national development priorities of the Government, but some of the leftist trade unions do not subscribe to this view. The fact is that the Companies Act, the Monopolies and Restrictive Trade Practices Act and the Foreign Exchange Regulation Act regulate the entry and operation of MNCs. The same set of labour laws that are applicable to domestic firms apply to MNCs also. Although there is no separate permanent machinery for consultation with workers' and employers' organizations concerning the operation of MNCs, the existing tripartite machinery and the parliamentary fora are used for discussion on MNCs. The employers and workers in MNCs are represented by their respective central organizations; there are no exclusive organizations of workers and employers for MNCs.

Privatization

Economic liberalization in India since 1991 resulted in a major policy shift permitting the entry of the private sector into areas hitherto reserved exclusively for the public sector. This ensured a measure of competition for public sector monopolies in sectors like transportation, electricity, oil, telecoms, banking and insurance, to mention a few. Along with this covert route, the government has also resorted to disinvestment of less than 20 per cent of the shares of about 50 central public sector enterprises (PSEs). During the 1990s, employees were also offered shares in central public sector undertakings as a sop to ease workers' opposition to privatization. Though some national federations of trade unions opposed the move initially, workers and several unions in some profit-earning sectors demanded the earmarking of a higher proportion (up to 26 per cent) of shares for them. When the shares of several of the enterprises began to decline, the initial euphoria among workers subsided.

The Government of India appointed a Disinvestment Commission in 1996–97 and referred about a quarter of the companies under its control to advise it on their privatization. The Disinvestment Commission made recommendations, from time to time, individually on about 50 central public sector undertakings. Successive governments did not, however, follow up on any of the major recommendations. Scooters India, the first PSE at the national level to have been considered for privatization in the early 1990s did not find a suitor because of its persistent losses, and because the government was not willing to allow any reduction in its workforce. It has since diversified its product portfolio and begun to earn profits. In 1999 the Disinvestment Commission was replaced by the Disinvestment Department.

The main elements of central government policy on PSEs announced in the Union Budget for 2000–01 include the following: (a) restructuring and reviving the potentially viable PSEs, (b) closing down PSEs which cannot be revived, (c) bringing down government equity in all non-strategic PSEs to 26 per cent or lower, if necessary, and (d) fully protecting the interests of workers. There are dissensions within the ruling alliance and among the opposition parties, with left parties vehemently opposing disinvestments. In regard to protecting the interests of workers, even if retrenchment is ruled out, downsizing cannot be, particularly because PSEs themselves have started downsizing. The Steel Authority of India which has proposals to reduce its workforce by at least one-third has even created a board-level position – Director, Restructuring – and engaged international consulting firms to consider shedding non-core activities and divest non-performing assets. The only major means of dealing with labour redundancy seems to be through voluntary retirements that give four times the normal retrenchment compensation and a promise of employee stock options in disinvested firms. Recent trends in the stockmarket have dampened the euphoria about employee stock options even in the ICE – information, communication and entertainment – industries.

In January 2000, Modern Bakeries became the first central public sector undertaking to have been sold to Hindustan Lever, the Indian subsidiary of Anglo-Dutch multinational, Unilever. Another 17 units were listed for privatization. These include Maruti Udyog where Suzuki is the majority partner, Air India, and some companies in the petroleum sector. Besides several trade unions, there is, however, consider resistance within the alliance partners in the Government as well as from several opposition parties. Among the trade unions, it is only the Indian National Trade Union Congress which is not showing any ideological opposition to

Table 12.1 Disinvestment in public sector undertakings

Year	Target (Rs crore)	Achievement (Rs crore)
1991–92	2500	3038
1992–93	2500	1913
1993–94	3500	Nil
1994–95	4000	4843
1995–96	7000	362
1996–97	5000	380
1997–98	4800	902
1998–99	5000	5371
1999–2000	10000	1479*

* This figure is to 31 December 1999.
Source: *Economic Survey (1999–2000)*, pp. 119ff.

privatization as such, even though the Congress Party with which it has strong links and which initiated the economic liberalization has began to question the stripping of national assets. It is seen from Table 12.1 that disinvestments have picked up in 1999–2000.

The finances of some state governments being in a much worse condition than the others and the central government, they had privatized some of the units owned and controlled by them before the central government could do so in respect of units under their ownership and control. Some states like West Bengal, which continue to be under left-front rule, found it more difficult to mobilize public opinion to sell state-level public enterprises like the Great Eastern Hotel in Calcutta. In the largest state of Uttar Pradesh, privatization of a cement factory, accomplished with considerable bloodshed and the killing of several people in police firing, had to be reversed in the wake of political instability that resulted in a change in government. The same state had taken a tough stand with striking power-men in early January 2000 over the privatization of electricity distribution because of its dependence on international borrowings to carry on reforms.

The experience with several private airlines and passenger road transport services in some states was also not quite positive. Privatization of electricity transmission and distribution in Orissa was initially held as a success and a model. But when a massive cyclone damaged the electricity distribution system in most parts of the state, the private sector operators were unable to restore the infrastructure. In the neighbouring state of Andhra Pradesh, privatized distribution was followed by a phenomenal rise in electricity tariffs incurring the wrath of the public. Voltas Limited, an otherwise profitable private sector unit took the refrigeration-making units from Hyderabad Alwyn, the state-owned public sector giant in Andhra Pradesh. This adversely affected the company's bottom line and the Chief Executive who favoured the acquisition, had to quit. All things considered, there are doubts in certain quarters about the nature of private enterprise in India and its capacity to mobilize the needed resources and its ability to manage the privatized businesses effectively.

Social dialogue on privatization – particularly deliberations on the future of Indian Iron and Steel Company, nationalized textiles mills and Indian Drugs and Pharmaceuticals Limited – in the special tripartite committee meeting and tripartite industrial committee meetings only helped to delay hard decisions and in the process rendered the sickness of these units more chronic. It is not only trade unions but also the political and bureaucratic leadership which is against privatization. The absence of civil service and political reforms has ensured that neither privatization nor liberalization of public sector managements from political and bureaucratic controls has occurred.

The Government offered some of the sick units to trade unions for takeover but unions were not willing to go for the bait. There is a realization in the government that mere ownership change will not help the companies to improve performance. Therefore, it is now being proposed to disinvest more than 50 per cent of the shares and bring the management under effective control of private management. A section of the civil service and politicians themselves continue to resist it because ownership and control of public sector enterprises has provided political sustenance for them. There is a move to introduce the concept of golden share to mediate resistance. There has been a shift in the government's approach to privatization, yet there is no clear-cut policy. *Ad hocism* continues, with concern for achieving revenue targets from disinvestment. Perceived political risks and social tensions are causing the government to make haste only slowly.

Export processing zones

The Government has been impressed with the success of the Chinese Special Economic Zones (SEZs) and the Union Minister for Commerce is keen to replicate them in India. A section of Indian employers welcomed the idea in the hope that it would give them the right to hire and fire employees. However, in the wake of vociferous protests from trade unions, the Minister had to assure that such zones would not be exempt from the labour laws of the country. Going by past experience, however, even if the labour laws in such zones are not discriminatory, it would be difficult to unionize and secure protection of the kind that the organized labour force in the country enjoys.

In India the organized sector refers to the sector comprising public and private enterprises which are registered and come under the purview of any/some or several of the Acts and maintain annual accounts and balance sheets. Public enterprises include departmental and non-departmental enterprises in the public sector. Private organized enterprises include registered manufacturing, mining and quarrying, gas and water supply, private transport companies, registered schools, colleges and hospitals, and corporate trading activities and services. In contrast, the unorganized sector comprises, exclusively, private sector units which are, if at all, only marginally affected or regulated by labour or industrial laws. Enterprises in this sector are usually very small (typically employing less than 10 persons) and are rarely unionized; wages are low, working conditions harsh, and conditions are exploitative and often largely unsafe.

Skills training

India has, in effect, chosen to give more education at a higher price to a few who have already had more than average education, rather than work for sound education for all. The effect has been to have a trainable total workforce, and a flow of highly skilled people notoriously larger than the number of jobs available (Oxenham *et al.*, 1990). Low literacy, a vocational bias against technical skills and an occupational preference for non-production jobs, a mismatch between skills acquired and skills required, and a dearth of adequately/appropriately trained technical personnel for non-executive positions are among the major weaknesses that characterize the macro-level situation in India. An emphasis on general education and an insufficient interaction between industry and educational/training institutions has exacerbated these characteristics. Together their role in limiting technical and economic advancement of the country cannot be exaggerated (1994).

There is a need for attention to the following five aspects in regard to skills development:

Primary and vocational education

There is growing evidence in many countries throughout the world that unskilled people are losing jobs and livelihood and skilled people are able to keep their jobs and improve their earnings. With rapid changes in technology, markets and environment, skill obsolescence is growing. Employment is contingent on employability, and employability is contingent partly on skills and largely on attitude. The best insurance against job loss for individual employees and maintaining competitiveness for enterprises is to effectively nurture and nourish a culture of multi-skills in the place of mono-skills. This provides career resilience and career self-reliance. There is a need to encourage continuous modular education and training programmes for workers at all levels in small capsules but at more frequent intervals than is currently happening. Alongside, we should also create facilities and opportunities for interactive open learning systems, using multimedia technologies so that workers are encouraged to update and upgrade their skills at their own pace and time.

The Government should step up its investment in primary and vocational education and substantially revamp those areas. We may examine the relevance of the Swiss and the German models of vocational training, for example. We also need to take stock of the Scandinavian experience in proactive labour-market policies. Employers and their organizations should take a more active interest in shaping and supervising vocational education; both industry and academic and technical institutions should reach out for each other, develop partnerships, share resources and improve education and training.

A skills development fund

Several countries are focusing attention on skills development. Singapore and Malaysia, for instance, have set up skills development funds with contributions from employers as a percentage (usually 1.5%) of the wage bill. The collection and disbursement of funds under the scheme is made simple by using the existing network of commercial banks. Employers' organizations in India are skeptical about such funds being an additional levy on their wages bills which will go mainly into payment of salaries and other expenses of government employees who administer the fund. Kuruvilla and Chua (2000) highlight that in Singapore the skills development system was successful because it was part of an institutional context where several different institutions and practices worked together to increase workforce skills. Also, Singapore had several facilitating factors which are not common to India. Among them mention should be made of the following: the scheme was fairly narrowly focused, mainly on electronics, and secondly it had a technocratic government bureaucracy with a high performance-oriented culture.

India could perhaps consider social funding of education and skills development with community support, and corporate funding could be considered. It would be useful if some large industries and local/regional/apex chambers of commerce and employers' organizations considered adopting some training schools and run them as model institutions for possible replication. In areas where skill shortages are considered significant and substantial, even competing firms in the same industry could come together to establish industry/sector-based training institutions. Alongside, we need arrangements to strengthen the existing mechanisms for skills certification and accreditation, and an appropriate incentives structure to encourage marketable skills provision and productive job-creation.

Retraining

The National Renewal Fund (NRF) did not meet its objective. While over 80 000 took voluntary retirement utilizing NRF funds, barely 1000 people were retrained and 100 redeployed. The task is daunting. Several companies have clauses on redundancy and retraining, but retraining in an unemployment context would result in a situation whereby redundant workers in a family would compete with younger members entering the labour market for the first time. Therefore, retraining as a solution may give rise to a new problem or not provide a complete answer to the problem of unemployment. Even so, retraining should be a continuous process rather than an isolated event if human resource obsolescence due to technological, economic or other structural changes is to be minimized/averted.

Multiple skills/tasks

The one-person one-skill/task concept is giving way to one-person many skills/tasks. This is helpful to organizations in terms of better utilization of personnel in the workplace. It also helps individuals as an insurance against redundancy since at least some skills are expected to remain marketable and facilitate redeployment. It is rather naive to assume that everyone likes variety in their jobs. There is a need to clarify the purpose and orientation not only of workers, but also their supervisors and managers. Multiskilling requires a thorough review of work processes, assignments, appraisal and accountability. This is found particularly useful in assembly operations where sequential performance of operations involves built-in idle time. Several Indian companies have had a fair degree of success, while an overwhelmingly larger number of firms have had a contrary experience.

Skills passports

Companies may issue passports to employees listing the skills which the employee has learnt and acquired mastery and proficiency in. Such a passport is a method of recognition of skills acquired by the worker, and may help outplacement in cases of redundancy.

New approaches to work organization

Work and employment relations are in a state of flux and transformation. Among the factors that have contributed to this are advanced technology, the forces of liberalization, privatization and globalization, the pressures of mergers and acquisitions, the ever-increasing diversity in workplace and cross-cultural interactions, the influence of the media, increased consciousness about democratization and human/trade-union/worker rights, the growing experience and knowledge about the values, attitudes and behaviours of people, growing competition and a rise in the expectations of the new workforce, and flux and transformation in employment relations.

Declining labour intensity is making many people wonder whether, as in the past the tractor replaced the animal from the farm, whether the computer is now threatening to replace the man from the factory. During the twentieth century there was a shift from the Taylorist/Fordist systems of division of labour with mono skills and multiple job classifications and pyramidical levels/layers in the organization, to Toyotaism based on lean production with multiple tasks/skills, fewer job classifications and flatter, slimmer downsized/right-sized organizations. This trend has continued into the new millennium, but with a caveat: 'beyond lean production'. Routine, repetitive and middlemen activities are either automated, robotized or being reengineered through application of advanced information technology. The cost-plus attitude is being replaced by strategies based on cost-cutting and value addition to provide the competitive edge to companies. Work and workplace are being revamped continuously and rapidly, and in this scenario those who learn fast and are adept at adapting to the rapid changes will survive the threat of jobless growth and income insecurity.

The following shifts are taking place in employment relations:

- A shift in the power base from land/capital to knowledge.
- A shift in the use of human energy from muscle to mind, brawn to brain.

- Top-down to transparent two-way communication.
- Fear to favour to fairness as the basis of motivation.
- Leadership based on direction and control to one based on consensus and commitment.
- High expectations of both workers and organizations. Flexible/ adaptive work organization, continuous training in skills and attitudes, say and stake for employees becoming critical and indispensable.
- The new collectives at the workplace will not be trade unions, but semi-autonomous work teams. The new workplaces are replacing close supervision with leaders who are coaches, mentors and facilitators.
- Improving labour standards and fostering ethical behaviour are becoming business imperatives.

In addition, the forms of employment are becoming more diverse:

- Full-time employees.
- Part-time employees.
- Home workers.
- Casual/contingent/Badli/substitute workers.
- Contract workers.
- Migrant intra- and inter-country workers.
- Foreign workers.
- Tele-workers.
- Apprentices who are not workers but perform work while learning.
- Bonded labour.
- Child labour.
- Displaced workers.
- Disabled workers.
- Rationalized workers.
- Self-employed workers, etc.

The major concerns are:

- Already thin employment in the organized sector is becoming thinner. The number of jobs in the sector (28 million plus) are less than the number of people waiting for jobs in that sector as evidenced by registrations in the employment exchanges (around 40 million).
- The new forms of work appear to pay less and are less secure, more tedious and more dangerous.

- There is a shift away from lifelong employment to lifelong employability, and a consequential shift of emphasis from job security to income security.
- There is a need for proactive labour market policies for job search, skills inventory, matching demand and supply, information dissemination counselling, etc.

In the past, many large companies in the private and public sectors encouraged lifetime employment, but this is no longer so. In the past people stayed with a company in a single job/career. The new workforce is by and large aggressive, ambitious, footloose and loyal to themselves, with the result that today neither companies nor their employees are loyal to each other. Yet today's organizations are expected to become learning organizations and build a customer base, but with high employee turnover it is difficult for them to accumulate knowledge and become such learning organizations.

Mere commercial contracts encourage only on instrumental orientation. Companies which fail to create a psychological contract will not be able to involve workers and get their commitment to work creatively and to their full capacity. Workers who are not loyal to their employers cannot be expected to build a loyal customer base. Hence, an employment contract with weak attachment on either side can have potentially counterproductive effects if either party pursues and persists in seeking unfair advantage. In addition, contract work is increasing. Companies generally do not pay the same degree of attention to the recruitment, training and safety, wages and productivity of contract workers as they do with their regular employees. Often, when courts order companies to regularize contract workers, they start raising questions and qualifications, experience and suitability. But such arguments are often not credible because they were not raised when the particular contract labour was hired in the first instance.

There have been questions about whether a contract is for work or of work. In either case the ultimate tests are the business and dependency tests. On both counts companies usually stand exposed because of the way in which contracts are structured/implemented. Also, when contracting out becomes a source of exploitation of cheap labour and an aggrieved party goes to court, such arrangements are questioned by the judiciary. There are also instances of new employment contracts based on notions of a contingent workforce, a crude reminder of the earlier, old putting-out system. Trades distributed raw materials and tools to home workers, who would undertake the assigned tasks and bring back the finished products to the traders. Today's producers without factories are akin to such traders. In some cases workers/trade union leaders are turning out to be major labour contractors. In yet other cases, production decisions are left entirely to workers who organize themselves without a foreman or a contractors.

Companies need to draw a line on how much temporary help and contract employment to use and what implicit or explicit continuity to offer employees. The short-term economic and long-term economic and social consequences of new employment practices need to be evaluated carefully. Unfortunately in India we do not have empirical and longitudinal databases and hence we tend to go by hunch, hindsight and anecdotal case histories/experiences. In view of that, it is important to formulate well-though-out and comprehensive approaches to outsourcing and job security.

Enterprise flexibility

Firms maintain flexibility through organizing their production in diverse ways and through strategic decisions such as the following:

- Parallel production;
- Outsourcing;
- Lease licence manufacturing;
- Franchising;
- Employment of a contingent workforce;
- Shifting workforce from contract of employment/work to contract for work, etc.

Flexible practices in labour utilization should not increase rigidity for labour. According to the labour law in several Southeast Asian countries, the rights to recruit, reward, transfer, motivate, assign work and adjust the workforce are considered managerial rights. In India, however, these are the subject of collective bargaining. The applicability of several labour laws increased with the increase in the size of employment. This serves as a disincentive for managements to reduce employment below certain threshold limits to come out of the purview of labour laws. This has, along with other factors, contributed significantly to the increase in capital intensity and a decrease in labour intensity in several industries.

Firms seek labour flexibility on one or more of the following counts:

- Numerical flexibility (size of workforce);
- Skill flexibility (composition of workforce);
- Functional flexibility (job enrichment/job enlargement);

- Locational flexibility (transfer/mobility);
- Time flexibility (flexi-time);
- Pay flexibility (flexi-pay).

International labour standards

As a founder member and as one of the ten countries of 'chief industrial importance', India holds a non-elective seat in the governing body of the ILO. India has ratified 38 of the 181 conventions. The ILO has influenced India and India has influenced the work of the ILO. A tripartite committee on ILO conventions overseas aspects relating to international labour standards, proposals concerning new conventions/ratification of old conventions, compliance of provisions in the ratified conventions, and so forth.

As seen from Table 12.2, the Constitution of India upholds all the fundamental principles envisaged in the seven 'core' international labour standards. India has also ratified three of the eight core conventions that constitute the fundamental principles, and is also likely to ratify, very soon, two more of the remaining five core conventions: Convention no. 105 concerning abolition of forced labour, and Convention no. 182 concerning immediate action to end the worst forms of child labour.

Table 12.2 Components of the social clause and Indian legislation

Social clause aspect	**Indian constitution/legislation**
Freedom of Association and Right to Collective Bargaining (Convention nos. 87 and 98 respectively)	Freedom of Association is guaranteed as a fundamental right in the Indian Constitution. The Trade Union Act 1926 meets with part of the objectives of Convention nos. 87 and 98. Both conventions are, however, not ratified by the Government of India
Forced Labour Convention 1930, and Abolition of Forced Labour Convention 1953 (Convention nos. 29 and 105). It provides for progressive abolition of forced labour in all its forms. Convention no. 182 concerns Immediate Action to end Worst Forms of Child Labour	Article 23 of the Constitution and the Bonded Labour System (Abolition) Act 1976. India has ratified Convention no. 29, not 105. India is moving towards ratifying Convention nos. 105 and 182

Table 12.2 (Continued)

Social clause aspect	Indian constitution/legislation
Equal Remuneration Convention 1951 (Convention no. 100). Its purpose is to eliminate sex-based discrimination in remuneration and provide for equal remuneration, to both men and women, for work of equal value. The four underlying bases for determination of work of equal value are: skills, efforts, responsibility and working conditions	The Constitution upholds the principle of equality between men and women. The Equal Remuneration Act 1926 seeks to provide for equal remuneration between men and women
Discrimination (Employment and Occupation) Convention 1958 (Convention no. 111). It covers any discrimination, exclusion or preference '. . . which has the effect of nullifying or impairing equality of treatment' and which can be the result of not only legislation but also of existing factual positions or practices	The Constitution upholds equality, denounces discrimination and encourages preferential treatment to disadvantaged groups in society. Convention no. 111 is ratified
Minimum Age Convention 1973 (Convention no. 138). It provides that the minimum age for employment should ordinarily be 15 and raised to 18 in dangerous occupations	The Child Labour (Prohibition and Regulation) Act 1986 bans employment of children below the age of 14. In several laws the minimum age is variously defined. Employment of children in certain heavy and hazardous industries (Schedule A, part 3) is prohibited by law and the government is taking steps to strictly enforce it. Several court judgments under public interest litigation actively support the prohibition and regulation of child labour

India is actively considering the ratification of Convention no. 105 concerning the abolition of forced labour. India has not ratified Convention nos. 87 and 98 concerning freedom of association and the right to collective bargaining due to 'technical difficulties' involving trade union rights for civil servants. This, however, was not a major hurdle because while ratifying the concerned conventions, the government can always exempt certain services. The real intention seems to be, as former Chief Labour Commissioner of the Government of India, Nath (1997) opines:

> These conventions have not been ratified by India because the policy of the government has been to restrict freedom of association to only manual workers (by defining them as Workmen) and exclude supervisory and managerial workers... The other interest of the government is not to allow the right of collective bargaining even to industrial workers in government's departmental undertakings like the Railways, Post and Telecommunications, Central Public Works Department, etc. The government, on the basis of pay commission's recommendations and not through collective bargaining, decides their pay, etc.

Several conferences organized or cosponsored by the Union Ministry of Labour, Government of India, including the one held in Mussoorie in the summer of 1998, discussed the matter and deferred the decision. Both employers and the government are one in this regard.

There are certain restrictions on the formation of federations in the telecoms sector. In 1981 the right to collective bargaining was taken away from insurance sector workers. This was done by the Parliament when it was found that collusive arrangements between the unions and employers (public sector) has either ignored or undermined the interests of policy-holders. Since then insurance workers engage in consultations, but their pay revisions are notified unilaterally by the concerned department in the Government.

India's record in respecting freedom of association and collective bargaining is much better than many other countries in the region, which have ratified either or both. Mere ratification cannot guarantee the achievement of the underlying objectives. Ratification should be backed up by a strong political will, a legislative framework and effective vigilance by the other social partners and the constituents and partners in civil society such as non-governmental organizations, consumer groups and so on. These considerations, however, cannot be a justification for non-ratification. India has ratified Convention 11 concerning the Right of Association for agricultural workers 1921, way back in 1923. The Union Ministry of Labour in cooperation with the National Labour Institute set up by it in the 1970s organized several workshops and rural camps to facilitate the organization of rural workers. Still, given the huge social inequities both in terms of caste and land distribution, the density of unionization among rural workers is negligible.

India has a long way to go before it can comply with core labour standards. 92 per cent of its workforce who are in the informal sector would be content at the end of the millennium even if the minimum standards in national legislation – not international labour standards – were the maximum. In the slate industry in one of the Indian states, the trade union leader, an influential policy party and district government officials wanted to help the workers secure minimum wages. Duly encouraged, the workers went on strike for three days but the employers, who had multiple businesses, did not feel the heat and therefore did not show any eagerness to end the strike. On the fourth day the workers agreed to return to work, but the employers agreed to take them back only after further reducing their wages as a token of punishment for their (the workers') temerity in taking them on (the employers). In another case, a young and committed Indian Administrative Services Officer in the State Labour Ministry was not amused with the flourishing dollar-earning tobacco merchants getting away with paying less than minimum wages to their poor, non-unionized labour. He computed the differences in the wages payable according to the law and wages actually paid to the workers, and got the dues paid to the workers under supervision by him and his staff. After the payments were made, he (the labour commissioner) and his staff left. Soon thereafter the hired *goondas* of the employers emerged to take back the money from the workers with a warning that if they reported the matter they would lose their jobs and their families could be in danger. In a third case, a government official told the author that he personally got a bonded labourer freed from a landlord in a village in Bihar. The person so freed was given an amount of money with which he bought stock. The buffaloes fell sick and died. The person wanted to seek employment in the village to eke out a living, but the landlord concerned warned the villagers not to take the person into employment, as he was a troublemaker. Finally the person had to return to bondage, this time on even harsher conditions of employment. The above-cited examples are not exceptions; they are common occurrences in many remote and rural areas. In the absence of a proper supervisory mechanism that fully understands and deals with the ground-level realities, efforts to improve labour standards could lead to their deterioration.

In contrast, there are some highly organized and unionized sectors where collectively determined labour standards exceed the national and international labour standards in some aspects. The private sector steel giant, Tata Iron and Steel Company (TISCO), had, during the 1920s and 1930s, introduced a number of voluntary welfare measures long before such measures were incorporated either in national legislation or international labour standards.

In the past some of the legislative initiatives based on ILO conventions have had the opposite effect to that intended. Measures to safeguard the interests of women through maternity benefits and restriction of employment of women in nightshifts in factories and in underground mining resulted in many employers hesitating to employ women and led to a drastic decline in the employment of women in textiles and mines. These measures should not, however, lead to discontinuance of maternity benefits. Introducing paternity leave would perhaps restore a balance and remove the incentive for employers to employ only males. The ILO Convention of Nightshift Work had been revised in the 1990s to enable the employment of women in nightshifts. The Government of India is pursuing, rather half-heartedly, changes to some of the labour legislations, and the employing ministries have apparently been pressing for some of these changes, including changes in the Factories Act to permit employment of women in nightshifts, particularly in electronic units and export zones.

India is actively pursuing 14 projects to eradicate child labour in hazardous industries by 2002. The All India Organization of Employers has undertaken a project in Jalandhar (Punjab) to have an agreement similar to the one in Sialkot, Pakistan, concerning the abolition of child labour in the manufacture of sports goods.

India has also advocated the promotion of labour standards within the framework of the ILO Constitution. It has consistently opposed proposals to link labour standards and trade through 'social labeling', etc., and the non-aligned countries' summit organized by the Labour Minister of India at New Delhi in 1995 adopted a resolution to this effect. India also played an active role in Seattle in 1999 to prevent linking trade with labour and environmental issues. All the three social partners – Government, employers' organizations and national trade union centres belonging to different persuasions – are all united against the linkage of international standards with trade (for statements of different social partners see IIRA/FES, 1996) for familiar reasons that are articulated in most developing countries throughout the world: commitment to the ILO's pillars of voluntarism, tripartism and free choice of social partners. Mandatory imposition of labour standards, by whatever name they may be called, contravenes Article 19(3) of the ILO Convention. All the social partners in India stand for upgrading labour standards, but they are against linkages within the context of the WTO as it is currently constituted since they suspect that the linkage is aimed at putting artificial barriers against competition. Concern for improving labour standards should be more holistic and encompass the entire working class rather than the microscopic minority engaged in production for exports.

Under Article 236 of the Constitution of India, labour is in the Concurrent List. Since liberalization there has been a tendency on the part of some state governments to pursue competitive labour policies with a view to attract investments and create jobs. The Constitution Review Committee appointed by the Government of India in 2000 should examine the need to look at this aspect with a view to considering the need for a unified law on matters which fall within the ambit of the eight core labour standards under the 1998 ILO Declaration on Fundamental Principles.

A national consultation on international labour standards where several non-governmental organizations and national trade union centres participated made a proposal with the following four components (CEC, 1996): (1) reject the labour rights – WTO linkage; (2) uphold the principles of universal labour rights and the need for evolving structures to monitor the enforcement of labour rights; (3) set up a UN Labour Rights Commission as an alternative; and (4) establish, at the national level, a powerful National Labour Rights Commission to monitor and enforce labour rights. A final point is that the 1998 ILO Declaration on Fundamental Principles and Rights at Work and its Follow-up substantially addresses the central issue. So long as the ILO takes time to make its instruments yield the desired results and the resistance to the WTO continues to build up, initiatives at the industry level through social labelling and at the community level through consumer boycotts will apply the kind of pressure that intergovernmental and international organizations are ill-equipped to exert.

During the 1990s several non-governmental organizations (NGOs) and corporations have made efforts to communicate that their products and companies do not involve child labour, using what is now known as 'social labelling' (US Department of Labour, 1987). In addition to creating a child labour-free production environment, some have even provided or funded alternatives for child labourers such as schools, employment for parents, and so on. Social labelling is mainly prevalent in the export segment of four industries, viz., hand-knotted carpets, garments, leather products and tea. The main child labour labelling programmes for carpets are: (a) Rugmark; (b) Kaleen; (c) STEP; and (d) Care & Fair. Rugmark and Kaleen are product labels fixed to individual carpets. STEP and Care & Fair are company certification programmes. Participating companies use the labels for advertising and marketing purposes, but not to individual carpets.

The Rugmark Foundation was established in India in 1994 and expanded to Nepal in 1995. It is a private, voluntary certification programme providing market-driven incentives for carpet manufacturers to produce without child labour. Under the programme, individuals and companies in the carpet industry are organized to stop the use of child labour, and an independent, professional and international credible monitoring and certification system for carpets manufactured without use of child labour was put in place. The affected children have been rehabilitated through education. As of May 1997, the Indian Rugmark Foundation had certified the export of about 636 000 labelled carpets from 164 licensed producers/exporters. The programme is monitored and enforced through an elaborate system of license approval, random inspections and carpet tracking. It conducted 1754 unannounced visits to licensees' facilities, found 143 children illegally working at looms and put them in two schools that it has helped establish.

The Carpet Export Promotion Council (CEPC) established by the Ministry of Textiles, Government of India, as a quasi-government body, established the Kaleen labelling programme in India in June 1995. It is funded partly through government grants and partly through subscription income from industry members. The members of CEPC commit themselves to the elimination of child labour and register looms of exporters. An independent agency conducts inspections and the Development Commissioner conducts periodic reviews. All registered exporters pay a fee for child welfare activities. As of May 1997, 572 000 Kaleen labels had been issued to 219 CEPC members, roughly 10 per cent of the total membership.

The STEP and Care & Fair programmes initiated by Swiss and German NGOs in the carpet trade, respectively, are essentially company certification programmes. Through the programme the suppliers in India and elsewhere are obliged to honour the terms of Demands, a Code of Conduct which includes elimination of child and bonded labourers, and to contribute to health and education programmes for carpet workers and their families.

Among the other labelling programmes, mention is made of Reebok International Ltd in the soccer ball industry. Through an agreement signed between the company and the exporters from Sialkot in Pakistan, in association with the Federation of International Football Association and the International Labour Organization, the soccer balls produced for Reebok bear the label, 'Guarantee: Manufactured without child labour'. The three major elements of the progrmame are (1) reorganizing all production in a new facility under one room where children below 15 years of age are prohibited to enter/perform work; (2) external monitoring to ensure that children are not entering the workplace and soccer-ball panels are leaving the factory for stitching; (3) support for the education/vocational training of children in the region. A similar agreement is being planned at Jalandhar in India.

The 'Fair Trade' labeling called the TRANSFAIR seal in the tea industry is another example. Its objectives are to ensure (1) the payment of a fair price, which covers production costs plus a fair trade premium that flows to the actual producers; (2) participation by the actual producers in decisions regarding the use of the fair trade premium; (3) provision for advance payment; (4) promotion of a long-term relationship between producers and importers. The avowed aim of 'fair-trade' labelling in tea is to

> promote producers in the 'Third World' who are at a disadvantage under present trading conditions and to help them achieve independence and equality. The additional price paid for the tea is used directly for the improvement of . . . the living conditions of tea pickers and plantation workers (for tea plantations) . . . the improvement of the smallholders' living conditions (for smallholders' cooperatives).

Social labelling exercises are still voluntary and thrive on actual or potential consumer boycott threats, but there is a growing awareness and acceptance of the labelling programmes. The costs are born by exporters and importers in several cases. Consumers in developed countries are also reportedly willing to pay a higher price for such labels, particularly under company certification programmes. There is also an ongoing effort to cover the subcontracting arrangements under the programmes.

In addition to labelling programmes, two kinds of codes of conduct are being popularized by the countries of the North. They concern (a) actual labour standards – what people are paid, how jobs are designed, the levels of safety, etc.; and (b) formal policies which purport to shape corporate conduct in certain ways. These are fundamental conditions which should apply across workplaces. At the national level the Clinton Administration in the USA announced a set of model business principles in May 1995 covering fair employment practices, including (a) avoidance of child and forced labour, (b) avoidance of discrimination based on race, gender, national origin or religious beliefs, and, (c) respect for the right of association and the right to organization and bargain effectively. These cover the eight core conventions under the 'Fundamental principles of international labour standards'. Ironically, the USA has ratified only one of the seven core conventions of the ILO.

Recent developments in industrial relations

The trend towards globalization has tilted the power balance in favour of employers; technology has reduced labour intensity but also diminished labour's control over jobs. The decline in employment in the organized sector has meant a decline in union density. Of course, there is considerable potential for an increase in union density in the unorganized sector.

Collective bargaining

Decentralized collective bargaining has reduced the ability of unions to disrupt and paralyze work and thus reduced their bargaining strength. Many of the cooperative agreements have meant assertion of managerial rights and forfeiture of accumulated rights of workers, even if some of them were in the nature of restrictive and wasteful practices. In sick industries unions have conceded agreements which provided for reductions in manpower, reductions/freezes of wages and benefits, and suspension of trade union rights such as the right to strike, for a certain period. In the sick public industries there had been no wage revision since 1992. In the PSEs the current trend is to sign agreements for 10 years as opposed to 3 to 5 year agreements during the 1970s to 1990s. The concept of parity among PSEs has been broken. Even in the Civil Service, the Eleventh Finance Commission proposed that in future there will be no need for periodic pay revision because employees are covered by full neutralization against rises in the cost of living.

Workers' participation in management

There has been much rhetoric about participation since Independence, but in reality it remains elusive to this day. Walker (1975) observed, based on his global survey, that the world over it was confined largely to tea towels and toilets. One wonders whether in India it has extended even this far (Venkata Ratnam, 1992). The Government has introduced many schemes of workers' participation in management since the 1940s, amended the Constitution of India to incorporate the subject as one of the directive principles of state policy, and took steps to encourage workers' participation in equity in the 1980s and to introduce stock options for employees in the 1990s.

The Government's discontentment with the implementation of voluntary efforts resulted in the convening of a national seminar and the subsequent introduction of a bill in the Rajya Sabha in 1990 to introduce workers' participation at all three levels – board, plant and shopfloor – through legislation. The bill is still (July 2000) to be taken up for discussion. Issues concerning the mode of representation, scope of the forums, levels of participation, coverage of the schemes, voluntary nature of the schemes, and so on, still remain unresolved.

Further, there is ambivalence about the true purpose and role of participation. Is it limited to giving workers and unions a say in management so that they, particularly unions, can pursue the sectarian interests of their members more vigorously? To what extent would unions accept the responsibility that goes with participation? Employers are sceptical about whether unions can and would like to accept responsibility. Unions, too, are generally wary of any scheme of participation that thrusts upon them more responsibility than effective power. Unions generally want the process to start from the board level and percolate down to the shopfloor. Employers want to start it, gradually, and with the shopfloor. There are only arguments, and hardly any agreement.

Consultative processes take time, and managers are wary of the costs and consequences of delays in decision-making. It is realized that the greatest obstacle in the way of labour management consultation could be the rapidly changing environment (see also APO, 1991). This is compounded by the adversarial mode of labour management relations, with a lack of mutual trust and, consequently, information-sharing, particularly on finances. This has the result that even Scanlon-type profit-sharing incentive schemes too become non-starters. However, many companies – including those in traditional manufacturing sectors – have started introducing stock options for their employees.

Aligning labour policy

A labour-market policy should balance the interests of the social partners. If one were to merely look at the labour market, one would be concerned more by considerations of illiteracy, poverty and unemployment. When one looks at the product markets, one would notice the trend towards increased competition, and the pressure to cut costs and/or add value. Workers/unions harp on labour markets and employers focus on product markets. The labour market impact is a function of the interaction between product market and labour market forces. In quite a few situations industry and product market characteristics strengthen employers and weaken or make the unions vulnerable. In some others, both will be left with limited choices/options and become defenceless, vulnerable or endangered. In the first set of industries employers can marginalize the power of trade unions through parallel production, outsourcing, and so forth. In the latter set of industries, input costs might be rising and output prices falling, squeezing both labour and management and leaving both with limited options. They are damned if they cooperate; they are damned if they do not.

Labour management partnerships in some cases can become suicide pacts in do-or-die situations. Given the lack of labour market flexibility, such dilemmas lead to dithering behaviour and preempt the potential for timely and proactive responses from both labour and management. It must, however, be said that labour market flexibility is *a* factor, not *the* factor. Poor physical and weak social infrastructures may have added to the problems in India's labour market. Input and infrastructure bottlenecks reduce job and income-earning opportunities. Recent experience of India's export performance in the post-liberalization period reveals that the industries which did well – software, gems and jewellery and clothing – were the ones which have competitive advantages in labour and skills. They were also – unlike the engineering industry, which is facing a serious recession domestically – less dependent on inefficient domestic input producers and infrastructure providers. At the minimum, what investors, particularly foreign investors, expect is:

- A clear enunciation of the rights and responsibilities of employers and workers/unions;
- An unambiguous and easily understandable legal and institutional framework;
- Predictable arrangements concerning union recognition, collective bargaining, skills development, flexibility and workforce adjustment;
- Well-defined, clear-cut and time-bound procedures for grievance redressal; and
- An administrative and judicial system that can be trusted for its transparency, integrity, expedience, efficiency and accountability.

From the workers'/unions' point of view, labour policy and labour law should stress support for:

- The observance of a minimal number of core/basic labour standards;
- Free trade unions and collective bargaining;
- Workforce institutions capable of internalizing the enforcement of labour standards/government regulations and effecting changes at the micro level smoothly;
- Investment in education and training;
- Bringing the entire labour force under the purview of a minimal, but effective – rational and rationalized – regulatory/administrative framework;

- Proactive labour market policies that provide for building skills/competencies reducing/eliminating the existing mismatch between acquired and required skills, facilitating information and counselling facilities for employment; and
- A culture of non-interference by one party in the affairs of the other.

Cooperative relations between labour and management are necessary to align the interests of individual employees with those of the organization through creative and innovative solutions to vexing human problems by integrated win–win approaches that take into account not only the sectional gains of labour and management, but also the other interest groups in society. Firms' abilities to adjust to new business pressures will be contingent on the quality of employment relations in the workplace. Workplace industrial relations systems should facilitate change, promote flexibility and prepare the workforce to be able, adaptive and attuned to respond to the emerging challenges.

Industrial relations

Workers have become more instrumental in their orientation, and the strike as a weapon is becoming blunted. Trade unions' objectives remain the same even in the new environment, but their means of action to achieve them cannot be the same as in the past.

The response of employers and trade union to industrial relations is often driven by the context in which the interface between the two social partners occurs. It is possible to discern at least five distinct scenarios where the balance of power seems to swing in either's favour, influencing the possible outcomes. Based on this typology it should be possible to read the signals early enough and predict the degree of choice and discretion that parties may have in influencing industrial relations outcomes.

Scenario 1

Input costs are rising and output prices declining; cash-flow problems limit opportunities for further investment, modernization and growth; productivity improvement is imperative to maintain existing levels of profitability. Major constraints persist in continuing to do more with limited resources over a longer period. The resultant squeeze in employment, wages, benefits, etc., creates a sense of helplessness and causes tensions/conflict in human resources/industrial relations. Such companies are too numerous.

Scenario 2

Several companies have, over a period of time, multiplied business turnover and profitability through (a) setting up parallel production facilities, (b) loan/lease license arrangements, (c) subcontracting, (d) franchising, etc. This has led to production/services becoming more lucrative for employers. As output/services in a unit become less critical, employers seem to develop a sense of independence and force unions to come to terms with management proposals. Several multinationals (Bata, Hindustan Lever and a host of pharmaceutical companies, for instance) which have conceded whatever unions asked in a cost-plus situation are now beginning to take (undue?) advantage of the increased vulnerability of unions in the liberalized environment since the late 1980s.

Scenario 3

When companies become bankrupt due to mismanagement etc., employers have to yield to union pressure to give up control and may be forced to accept employee buy-outs by worker cooperatives, for example. In quite a few cases (Kamani Tubes and Central Jute Mills, for instance) workers are able to revive sick units and sustain their profitability with support from professional management.

Scenario 4

When economic, structural, technological and other changes make enterprises bankrupt, trade unions are agreeing to several concessions of the following type to revive the unit and save threatened jobs: (a) downsizing, including retrenchment of a section of the workforce; (b) wage and benefit cuts and freezes; (c) a freeze on cost-of-living and other allowances; (d) suspension of trade union rights for collective bargaining and industrial action for a certain period; and (e) agreeing to changes in work norms and work practices and greater managerial discretion in maintaining discipline, production and productivity, etc. In almost all the companies – both public and private sector – whose cases are referred to the Board of Industrial and Financial Reconstruction (BIFR) for revival, these concessions have become the norm rather than the exception. The latest in the series is the multi-unit public sector firm, Indian Drugs and Pharmaceuticals Ltd.

Scenario 5

When neither the trade union nor the employer seems to show adequate concern and sensitivity to workers' interests, workers are seen to ignore both employers and union and take charge of the situation. They neither prefer a strike nor are they prepared to face a lockout. They simply occupy the plant and run it to keep their jobs and earn their livelihood. The most recent such case is that of the Kanoria Jute Mills in Calcutta.

A charitable interpretation of the current industrial-relations scene in India would be to call it very complex. A critical view could be that it is chaotic. The dynamic change process warrants a fresh look at the institutional and legal framework as also the roles of the principal actors. Changes in industrialization strategies call for changes in labour management policies too, to obtain the requisite balance between the social system in our organizations and the technical systems for the performance of organizations towards fulfilling the purposes for which they are created.

Workplace changes point to a reduction in manufacturing employment, downsizing as right-sizing, the ascendancy of employer rights, and increasing vulnerability of unions due to technological, economic, structural and other changes. In employment relations various shifts are discernible. The shift to a market economy being decisive and irreversible at least in the foreseeable future, organizations will need to do more with limited resources to be viable, competitive, productive and profitable. The focus will be on productivity, profitability and quality improvement.

Roles of social partners

Working together should be the spirit of social partners with the following motto: 'Together we aspire, together we achieve, and together we share with all our stakeholders.' Towards this end, the following agenda is suggested as possible roles for the three key social partners.

Government

1 Government should conduct itself in the role of a facilitator rather than a regulator and controller.
2 It should rationalize its role in regard to the organized sector and strengthen its role in the unorganized sector.
3 It should accept responsibility for building a supportive social and physical infrastructure and a legal and institutional framework.
4 It should put in place proactive labour-market policies including skills provision, supply–demand matching, counselling, a social safety net, etc.

5 It should consult the other social partners in formulating and implementing social and economic development policies in the true spirit of ILO principles of social dialogue, and widen the social dialogue to make it more inclusive.
6 The focus should be shifted from dispute-resolution to promotion of sound labour-management relations.
7 It should stress support for both efficiency and equity, and strengthen the labour ministry to ensure active participation of labour and other stakeholders for evolving policies for sustaining growth and competitive labour markets critical for ensuring job security and social protection.
8 It should widen the social dialogue.

Employers

1 There should be a basic change in the operative philosophy of management. They should move away from the concept of control to building policies and practices based on consensus and commitment, accepting workers and unions as partners in developing policies and solving problems which affect their organization.
2 The core values of labour management relations include fairness and equity, balance in the exercise of power and authority, trust and transparency, and promotion of policies which are conducive to collective well-being. Management being in the driver's seat must take the initiative and accept responsibility to ensure that these values are pursued in all earnestness.
3 Managers should view their roles as those of leaders, not policemen.
4 Management should focus on skills and attitude training for all employees, including the seniormost managers.
5 Management must solicit and implement ideas from rank and file employees and fully tap the expertise of its staff. This in turn enhances employees' self-worth and dignity and contributes to the success of the organization.

Trade unions

1 Trade unions should be full partners in the development process – both at the macro and micro levels.
2 Like managers, they should also change their operating philosophy and behaviour, paving the way for cooperative relations between

them and management harmonizing the concerns of their members with other stakeholders.

3 They should focus on increasing the size of the cake and seek a better slice; cooperate at the workplace level and confront at the bargaining table to get a fair slice.
4 Trade unions should have positive goals and roles rather than being mere managers of discontent. They should have a vision and create a desirable future for their members, themselves and society through well-thought-out goals that look at the issues in a holistic manner and address those concerning workers in a pragmatic manner.
5 Trade unions need to develop activities other than collective bargaining. Non-bargaining roles and social contributions will improve their image and identity within their membership and with other social partners in the wider society.
6 Even where trade unions feel compelled to oppose some policies like privatization, they should articulate their objections not merely from the point of view of the interests of their members, but from the implications for consumers and the community. Similarly, when they evaluate the impact of the changes, they should look not merely at the consequences of changes, but also consequences of non-change.
7 Trade unions should realize the consequences of their actions for members in the organized sector on others in the unorganized sector and develop strategies of approach that are not divisive and counter-productive, and do not result in increased social tensions.

References

Fallon, P.R. and Lucas R.E.B. (1991) 'The Impact of Changes in Job Security Regulations in India and Zimbabwe', *The World Bank Economic Review*, vol. 5(3), pp. 395–413.

Frensen, J. (1991) *Subcontracting and Inequality: The Case of Hindustan Lever in India*. Nijmegen: Third World Centre, Catholic University of Nijmegen.

Goyal, S.K. (1984) *Small Scale Sector and Big Business*. New Delhi: The Corporate Study Group of the Indian Institute of Public Administration.

Gupta, S.P. (1999) 'Trickle down Theory Revisited: The Role of Employment and Poverty', V B Singh Memorial Lecture, 41st Annual Conference of the Indian Society of Labour Economics, IGIDR, Mumbai, 18–20 November.

International Labour Organization (1999) *National Renewal Fund*. New Delhi: ILO.

Jalan, B. (1996) *India's Economic Policy: Preparing for the 21st Century*. New Delhi: Viking.

Indian National Trade Union Congress (INTUC) (1993) *25th INTUC Session Report: May 1988 to April 1993*. Cuttack: INTUC.

Johri, C.K. (1992) *Industrialism and Industrial Relations in India*. New Delhi: Oxford University Press.

Mathur, A. (1989) *Industrial Restructuring and Union Power: Micro-Economic Dimensions of Economic Restructuring and Industrial Relations in India*. New Delhi: ILO-ARTEP.

Nath, S. (1993) *Tripartism in India*. Bangkok: ILO APPOT Project (mimeo).

National Commission on Labour (Government of India) (1969) *Report of the National Commission on Labour*. New Delhi. Government of India.

Ramaswamy, E.A. (1988) *Worker Consciousness and Trade Union Response*. New Delhi: Oxford University Press.

Tulpule, B. (1992) 'New Industrial Policy, Employment and Structural Adjustment in India', *Indian Worker*, August.

United Nations Development Programme (UNDP) (1999) *Human Development Report*. New York: Oxford University Press.

Venkata Ratnam, C.S. (1991) *Unusual Collective Agreements*. New Delhi: Global Business Press.

Walker, K.F. (1975) 'Workers Participation in Management: An International Perspective', *IILS Bulletin* (Geneva).

The Workers Solidarity Centre against Job Losses and Closures (1989) *Report of the Workshop on Job Losses and Industrial Closures*. Seminar on Social Movements, Human Rights and the Law, 27–30 December, Bombay.

13 Globalization, National Culture and the Future of India

Dipak Basu

'Globalization' stands for the ultimate rights of multinational companies to allocate resources according to their own criteria of efficiency, which may or may not correspond to the national cultures of the countries in which they are operating. The idea is that an economic system can be successful if it corresponds to the culture or the philosophy of life of the host country, otherwise it will die out sooner or later. The purpose of this chapter is to examine whether the globalization process corresponds to the basic philosophy of people's lives. An analysis of the Indian philosophy and the universal laws of nature suggests an alternative view of the world economic system. India's future will be determined by the interactions of the economic system and inherent characteristics of her national culture.

'Globalization' was achieved from the middle of the eighteenth century to the mid-twentieth century by multinational companies from Britain, France, Holland and Belgium, and lately during the early twentieth century by Japanese and German multinational companies in their respective empires. The British East India Company forced Indian farmers to produce indigo and opium to sell to China to purchase silk. Britain invested tax revenues from India to build railroads in the USA and industries in Brazil and Argentina. There was a huge mobility of labour, too. Indian and Chinese slaves (or indentured labourers) were employed in sugar plantations in the West Indies and Latin America, in railroad constructions in Africa, and in mining in Australia and Southeast Asia.

European multinational companies since the eighteenth century created the European empires. The Hudson Bay company in North America, the Dutch East India company in Indonesia, the British East India Company in China and India, the Cecil Rhodes mining company in Africa, and the Anglo-American oil company in the Middle East are some of the examples of how multinational companies have helped their respective governments to gain colonies which are now called the 'developing countries'. That type of globalization was halted only when, after the Second World War, the Soviet Union became the victorious power in Europe; a large number of colonies became independent, and an alternative and very successful strategy of economic development – the 'planned economy' – became the model for the newly-independent countries. When the power of the Soviet Union was at its zenith during the 1970s, the developing countries asserted themselves to implement a 'new international economic order', to reverse the adverse terms of trade of their exports maintained historically by the developed countries as the source of exploitation.

However, with the destruction of the Soviet Union the power of the so-called 'Group of 88' developing countries disappeared and it was possible for the developed countries to impose a very new international economic order through the all-powerful international organizations – the World Trade Organization, the World Bank and the International Monetary Fund. The so-called 'new order' or globalization means reestablishment of the old order that used to prevail before 1945. Thus, it should be obvious that a minority will gain, with a great loss for the majority in the developing countries, as the world in the twenty-first century has returned to the nineteenth century.

Multinational companies and national economies

During the colonial period, before the Second World War, the relationship between multinational companies and host governments was that between agents of the oppressor and the oppressed. The thesis put forward by Bukharin (1917), Hobson (1902) and Lenin (1917) was that multinational companies seek new markets to increase their profits and enlarge their scope of investments of surplus capital. The doctrine of free trade was used to justify that aim so that host countries would accept domination of their sovereign rights in exchange for economic growth, which might never occur. State powers were used to provide security for the interests of the multinational companies, which eventually resulted in imperial conquests. The interests of the multinational companies and their governments were identical in exploiting these colonized countries.

The effects of these investments on colonial economies were initially great social and economic upheavals, destruction of domestic industries and in some cases great famines and catastrophes. At the same time, some parts of the host countries, particularly in the coastal areas, benefited from their subscriptions to the global economy controlled by the great colonial powers.

Given that history, it was not surprising that after the Second World War most developing countries considered multinational investments as instruments of neo-colonialism and have tried to develop a series of regulations to restrict their powers and intrusion in their domestic economies. However, economic interests of the growing middle class in the developing countries, particularly in East Asia and Latin America, have colluded with the international interests of the multinational companies. The result is that gradually developing countries have accepted multinational companies as a source of investment and started creating conditions hospitable for them.

This tendency has become most prominent since the mid-1980s due to two major factors. First, the rapid developments of the East Asian countries including China are mainly due to foreign investments. The negotiations of the GATT (General Agreement on Trade and Tariff; now called the World Trade Organization) and instructions of the IMF and World Bank have created enormous pressures, both intellectual and economic, on developing countries to accept capitalism and globalization and to remove obstacles for foreign investments. Second, individual developed countries have also created economic pressure to facilitate entry of their multinational companies. As a result, developing countries are now competing for foreign investments and offering more and more attractive terms, which they do not even offer to their own domestic companies, to multinational companies.

There are a number of characteristics of multinational companies regarding their activities in the developing world. First, they operate as protégés in order to obtain protection from the host governments (Kudrle, 1991). Also, in several developing countries, Europeans and North Americans have intervened militarily over many decades to assert property rights of their multinational companies. Encouragement given by the United States in Chile in 1973 to destroy the government of Allende when he nationalized the copper mines; Anglo-French invasion of Egypt in 1956 when Nasser nationalized the Suez canal; or the coup organized by Britain and the United States in Iran in 1953 to destroy the government of Mossadeg when he nationalized the oil fields are some of the examples. Currently, developed countries withhold economic assistance and prohibit economic aid from multinational agencies if a country does not protect the property rights of multinationals or if it creates entry barriers.

Multinational companies also work as agents of the home countries. The US government forbids American companies to do business with Cuba, Libya, North Korea and India-Pakistan (the restrictions are for defence-related companies and research institutes, but not for all companies), and extends these rules to companies originating from some foreign countries if they want to do business in the USA. Furthermore, multinational companies, particularly in many developing countries, are considered to be symbols of the modern world, but penetrations of foreign and alien cultures can have devastating effects. Examples from Thailand and Latin America show the social and moral destruction caused by the foreign cultures spread by multinational business organizations (Sunkel, 1972; Cardosso and Faletto, 1979). Multinational companies in developing countries frequently manipulate not only government decisions but also the broader political process. Corruption and attempts to manipulate regulations on the environment, and tax evasions, normally include manipulations of the political system.

Recently, with the emergence of corporate sectors in developing countries, rivalry between multinational companies and the domestic corporate sector has become an important issue. In order to attract foreign investments host governments relax rules only for the multinational, and multinational companies can obtain sovereign guarantees on their rate of profits, relaxed regulations on labour management, taxation and international transactions, which are denied to domestic firms. Quite often host governments, under pressure from international financial organizations, the World Bank in particular, have denied government contracts to domestic firms. Recent modifications of international patent laws, originating from the World Trade Organization, particularly prohibit domestic firms to produce independently without any collaboration with multinational companies.

The problem is now acute in the pharmaceutical industries in particular. In India in 1995 the government under pressure from multinational companies closed down public sector pharmaceutical firms and forced other private sector pharmaceutical companies to collaborate with the multinational companies. In a number of studies (Cordosso and Faletto, 1979; Evans, 1979) of Latin American countries, it has been shown that in Brazil, Mexico and Argentina, multinational pharmaceutical companies have been successful in collaborations with the host governments of those countries in destroying highly sophisticated and profitable pharmaceutical industries and research organizations. These conflicts with domestic firms will be more acute when the laws of the World Trade Organization are implemented fully and most domestic firms in developing countries will have to forego their independent existence.

Multinational companies, particularly in developed countries, face conflicts with trade unions as well as the organized labour force who consider them as agents destroying the welfare state by reducing the national autonomy of the host governments. In the developing countries, however, where both the governments and trade union movements are weak, multinational companies face difficulties only in certain countries or in certain parts of the country. It is quite normal to expect that the host governments will use the force of the government to destroy any opposition to the multinational companies. Relaxed labour laws and environmental regulations in developing countries are some of the most important attractions for multinational companies in recent years. The globalization process is intensifying these tendencies to undermine labour and trade union rights in developing countries in their drive to attract foreign investments, when they are all trying to be like China where no labour or trade union rights exist.

The crucial question is whether the new system is sustainable and, if not, what system may emerge. If a new system of economic management is not rooted to the cultural values of the recipient country, or if because of its cultural characteristics the recipient country cannot adjust to the new reality, the new economic system will fail.

National culture and economic organizations

The sustainability of an economic system depends on how the economic organizations behave in response to the culture of the host society. Culture is a multidimensional construct comprising several layers of interrelated variables. As society and organizations are continuously evolving, there is no theory of culture valid at all times and locations. However, the core values of a society can be analysed by asking the following questions:

1 What are the relationship between human and nature in the context of the society?
2 What are the innate human natures embodied in the history of the society?
3 What are the modalities of human activity?
4 What is the basic relationship between one man to another in society?

These are the basic macro values shaping the culture of a society (Kroeber and Kluckhohn, 1952; Kluckhohn and Stodbeck, 1961; Miroshnik, 1997), the precursors and determinants of human behaviour and the decision-making process. There are also micro-values or perceived practices which are the roots of the organizational culture (Hofstede, 1991), and combinations of micro and macro values give rise to an organizational culture which is specific for a country. These are also certain codes of behaviour that influence the future and the expected behaviours of the members of society if they want to belong to the mainstream (Miroshnik, 1997). These vary from one society to another as the expectations of different societies as products of historical experiences, religious and moral values are different.

An organization is successful if it gives rise to an organizational culture, which is embodied in the national culture of the country in which the organization exists. Organizational culture emerges from some common assumptions about the organization, which the members share as a result of their experience as members of the organization. The national culture shapes the way these common assumptions emerge. These are reflected in the pattern of behaviours, expressed values and observed artifacts of the members of the organization, which are the superstructure of the psychology of the members. At the core lie assumptions; the next higher stage contains values of the organization.

Russian and Japanese cultures and organizations

Relationships between culture and the success or failure of its economic and social system are very intimate. I give two examples of how culture has shaped the destiny of the economic and social system in Russia and Japan.

In Russia, the Bolshevik party was a minority party in 1917 yet the revolution it started spread very quickly throughout the vast Czarist Empire. The reason was that the ideology of the revolution coincided with the philosophy of life in Russia – the 'Cossack' philosophy of life. Cossacks are the Russian serfs who had escaped from the tyranny and oppressions of their feudal landlords to the then (sixteenth century) outlying areas of Russia and Siberia to form their communes of free societies. They had to fight against all odds, hostile nature, Turks, Mongols, Arabs, and Russian and Polish landlords. As a result they developed a culture whose values are brevity, endurance, communal spirit and equality. The ideals of the Russian revolution corresponded to this philosophy of life and that was the reason for the success of the revolution in transforming a backward agricultural country into an industrial giant within a few decades. Although socialism was destroyed in 1991, the subsequent capitalism introduced in Russia has not taken root and has so far failed to increase social welfare, instead reducing the country to near destitution. Capitalism may not have any future in Russia, as it is not related to the Russian philosophy of life.

Japan has accepted capitalism introduced during the 1860s reform but has given it a completely different shape by remoulding it according to Japanese culture. Japanese culture is that of a rice-growing people. In a rice field it is essential to have cooperation between the members of the village to share water, seeds, harvest equipment and knowledge. It is not a job for an individualistic person. Culture in Japan is based on the principle that tradition, stability, cooperation and knowledge are much more important than self-seeking achievements (Hayashi, 1994; Hayes, 1981; Miroshnik, 1997; Morita, 1992).

It is the formalized organization or system which in Japan is revered as the standard framework for behaviour. Placed in circumstances divorced from these organizations or systems, the individual frequently seems to become repressed and inward-looking, since he does not normally have much occasion to refer back to universal principles or ethics that might govern individual behaviour and help him to act in accordance with his own autonomous judgment. In Japan, the individual is born not into the family, but into a local organization based on land and occupation.

Society, thus, is felt in terms of practical systems and organization. These are accepted as something given and self-evident, without clearly questioning their justification, and are perpetuated as norms that must be observed except under very exceptional circumstances. The maintenance of the status quo as seen in organizations and systems and the stability of society are the most important missions of the system. In the management of systems and organizations, it is considered better wherever possible to rely on precedent; people try as far as possible to avoid making exceptions or to step outside the framework of the system.

Few Japanese feel that their freedom of behaviour is restricted by existing systems or organizations; rather, they have always seen behaviour conforming to that framework as a natural and desirable way of life. The essence of the existing systems and organizations can only really be appreciated through the accumulation of experience; the system or organization is given precedence over the individual's personal creative will. The man, one might say, does not create the system; rather, the system creates the man. Emphasis is on the educational aspect whereby a person gradually accumulates experience within the system or organization, acquiring knowledge, skills and all kinds of responses while carrying out the role already assigned to him.

In Japan, the system or organization is not viewed primarily as one possible means whereby the independent individual may achieve his aims; on the contrary, it is considered that if he/she observes the system and joins in the organization, fulfilling his or her proper role, then achievement of the objective is guaranteed. The stress placed on precedent in running a system is likewise aimed at avoiding, as far as possible, management based on individual judgments. This tends to mean that not only selfish ambition but individual creativity is also suppressed. It means, too, that knowledge of past processes and precedents is valued more highly in the organization than flexible thinking.

Japanese society is one that has managed to combine a sense of community based on territorial ties and the traditions of the workplace community with industrialization and urbanization – something that is found, within East Asia, in neither Korea nor China. This relationship between the impersonal in the form of the corporate organization, and the personal, shows a marked difference from both China and Korea which attach overwhelming importance to personal exchanges and such strictly human relationships as personal loyalty.

If the liberation of the individual from the community is the criterion for modernization, Japan is in that sense far from being modern, and the individual Japanese has little interest from the outset in the pattern of development known as modernization; rather, he has always valued, and still values, a more Japanese type of identity.

In short, the Japanese system of industrial management which gave Japan its supreme economic status in recent times, is based primarily on Japanese culture. Stability is more important than quick results. Education comes from learning from others, which may mean total imitation. Japan used to imitate China and Korea before 1860; from 1860 to 1945 Japan imitated Britain, France and Germany; and after 1945 it is imitating the USA as a part of the learning process. However, learning does not end with imitations. Continuous improvements (*Kaizen*) and innovations are the next stages of learning, which have made Japan the winner in the world marketplace (Basu and Miroshnik, 2000).

Thus the success of Japan is based on Japanese national culture, and the present apparent inertia to change in the face of long recession of the economy since 1991 is also a part of the Japanese culture; that is, stability is more important than change. Although, Japan was introduced to western ideas and technology nearly a hundred years later than either India or China, it is Japan not China or India that has achieved successful industrialization. This is because it has absorbed and remoulded western capitalism to suit its own culture. Emphasis on organization is part of capitalistic culture; Japan has no 'universal' value system for individuals which can contradict the value system of the capitalism either. As a result, capitalism does not create any contradiction to the Japanese philosophy of life.

If the psychology of the members shaped by the national culture does not correspond to the values of the organization, the resultant organizational culture will be weak. The organization will ultimately be a failure. If a multinational company tries to impose its own values, which are alien to the national culture of the host country, contradiction emerges which destroys the efficiency of the organization. This is essentially the contradiction of globalization, which implies that there is no guarantee of efficiency in a globalized system.

Thus the question is, what are the philosophical values of globalization and are these compatible with the basic fabric of the cultures of recipient developing countries. If these are compatible, developing countries will be a part of the globalized system; otherwise, an alternative system will emerge.

The philosophy of globalization

The philosophical basis of the globalization process is the philosophy of capitalism, that is the utilitarianism of Bentham (1983), James Mill (1986), John Stuart Mill (1999) and other writers. The idea is that maximization of self-interests is the virtue and the rationalism. Individuals, while selfishly maximizing their own interests, maximize the combined social welfare of the society; the process was explained as the 'invisible hands of the market' by Adam Smith (1998). This virtue of self-interest is the motive force of capitalism and is the so-called 'Protestant ethic' (Weber, 1946, 1993). The ideas of modern-day economists who are the high priests of globalization are little different from their eighteenth-century predecessors.

According to Bentham (1983) 'what is good is pleasure or happiness... therefore one state of affairs is better than another if it involves a greater balance of pleasure over pain'. John Stuart Mill (1999) said 'pleasure is the only thing desired; therefore pleasure is the only thing desirable'. General happiness that results from these pursuits of pleasure is the effect but not the intention.

All human actions are based on self-preservation and self-interest. Selfishness is a virtue, which brings economic prosperity. In the pursuit of profit-maximization, producers allocate resources only to satisfy demands so as to use the most efficient production system and minimize costs. Consumers are satisfied to receive high-quality goods at the lowest cost. Because economic growth depends on acquisitive actions, self-serving behaviours are justified. To enhance economic growth, the state should intervene as little as possible restricting itself to the defense and judicial system leaving everything else to the spirit of free enterprise.

This doctrine of *laissez-faire* was propagated by the originators of modern Western economists, David Ricardo and J.B. Say, and further decorated, using mathematical tools, by Jevons, Marshall, Knight and Walras and very recently by Milton Friedman or Robert Lucas. The argument remains the same although society has changed and the 'perfect competition' as imagined by Adam Smith is not a reality in the days of a monopolistic market of large multinational corporations. The idea is that capitalism, left to itself, can recover from any crisis, and any public intervention can only make things worse. Thus public actions are nothing but distortions to the system, which must be minimized so that multinational companies can pursue their self-interests freely so as to maximize the interests of the world economy.

The efficiency of the market is to satisfy demand, which can only be created by people who can afford to create demand. Those who cannot are rejected by the market. As prices are determined by the monopolistic multinational companies, the number of people rejected by the system cannot be determined by the policies of the national governments. For a country with poverty, the number of these rejected people can go on increasing thus producing a growing army or so-called 'underclass' who exist in large numbers even in developed countries.

Although India and the developing world, along with the former-socialist countries since 1991, have adopted these ethics, whether society and its cultural basis in those countries can accept it is questionable. In India, for example, the ideas of Bentham were known to Raja Ram Mohan Roy, the father of the Indian renaissance in the nineteenth-century, but were ignored. They were mocked by the famous nineteenth-century writer Bankim Chandra Chatterjee as 'the Philosophy of the stomach' (Chatterjee, 1986). Whether the doctrine of selfishness as a virtue can be acceptable to Indian culture, which is based on renunciation and selfless work, is debatable. The issue is the same for other cultures as well.

Both China and India, although exposed to Western ideas and technology about one hundred years before Japan, still could not absorb Western technology or its motive force, while Japan became 'the developed' country of the world. The reason is that Japanese culture has always encouraged and promoted the best from any foreign land. The culture and philosophy of Japan readily accepts whatever is the best and dominant at any time, and remoulds it to suit its own culture. The culture of most countries does not accept foreign philosophy or ideas unless they are already related to its existing ideas. To understand that, we need to analyse some fundamental elements of Indian philosophy which has explained a historical cycle applicable universally.

Law of nature and the historical cycle in Indian philosophy

To understand Indian philosophy it is essential to understand the meaning of *Vedanta*, which is a kind of philosophical algebra covering all of India's philosophical trends. *Vedanta* philosophy has two major aspects: dualism and non-dualism. Dualism in the European philosophical system denotes a philosophical doctrine which believes in two opposite substances, material and spiritual, irreducible to each other to be the source of all being (Russell, 1946; Brodov, 1964). In Hinduism, dualism implies recognizing the existence of two worlds, the other world (the world of essence, *Brahman-Atman*) and this world (the world of phenomena, the universe, nature) (Vivekananda, Vol. V, 1946). Both matter and spirit are nature (matter); only the former is crude, the later fine. Matter and thought coexist. *Brahman-Atman* is a third element, of which both matter and thought are products. The external and internal natures are not two different things; they are really one. Nature is the sum-total of all phenomena. Nature means all that is (Vivekananda, Vol. VIII, 1946).

All *Vedantists* believe in cycles as the mainline of the development of all that is. The doctrine of cycles is as follows. All that exists in the universe is the result of the manifestations of primary matter called *akasha*. All forces acting in the universe whether they are vital forces or the forces of gravitation, attraction or repulsion are the result of the action of the primary force called *prana*. Acting on the amorphous *akasha*, *prana* constructs the multifarious universe (nature). There is also a cosmic reason or *mahat*. *Akasha*, *prana* and *mahat* all emerge from *Brahman*, the absolute.

Nature as a whole is bound by the law of causality and is in time and space. Matter is substance plus time, space and causation (Vivekananda, Vol. III, 1946). Space, time and cause and effect links should be viewed as the conditions of being of the concrete objects or phenomena of the moving universe or nature. Nature is *maya*, which means the form of all that is, of the infinite quantity of qualities. *Maya* or nature is a projection of the objectively existing *akasha*, the primary matter of the universe. *Maya* signifies recognition of the phenomena of the material world as they are and the basic fact that the very basis of our existence is contradictory, that development everywhere proceeds through profound contradiction (Vivekananda, Vol. IV, 1946).

The universe moves in cycles of waveforms. It reaches its zenith, then falls and remains in the hollow as it was. Human organizations are part of nature. Consequently, the history of human society is also subject to the objective law of cyclic development.

Man is an organic mixture of the materials of the universe; it is a synthesis of the universe or the universe on a small scale. *Vedantists*, according to Vivekananda, believe that nature (the universe) is a subject to a universal objective law whose action affects the body, the mind, the soul, all that is. Everything in nature is absolutely determined and law-governed. Law is not an idea or abstraction but an actual expression of real, recurring, stable and essential links between the phenomena of nature. All nature develops according to laws, manifesting itself through laws and acting according to them.

Man gradually established that his personal life was ultimately subject to the same universal law as all nature. That is the essence of the philosophy of life in India or the fundamental principle of Indian culture. The fundamental principle is the 'theory of *karma*', which says that each action sooner or later causes a certain effect. Everything in nature, from abstract thought to practical action is determined and directed by this law. Man sets himself the goal of freeing himself from the bondage of nature. The meaning of a man's life according to the Indian culture 'is the awareness of the soul to its bondage and its efforts to stand up and assert itself' (Romain Rolland, 1944).

According to the message of Krishna in the *Bhagawat Gita* this freedom can only be achieved by *karma yoga* (selfless works) and *gnana yoga* (pure knowledge) (Bhagawat Gita, ch. 3, verse 3; 1983). *Karma-yoga* recommends working for the sake of the work itself, not for the fruits of works. Work without pay, absence of attachment to the result, generally to the point of complete disregard for one's personal interest, complete selflessness, is *karma yoga*. This is essentially opposite to 'utilitarianism', which is the philosophy of globalization.

Arabindo Ghosh (1947), the noted philosopher, explained it further. The principal contradiction of human life is that between the individual and society, or aggregate. The essence of the ideal law of human development demands that the individual should harmonize his life with the life of the social aggregate. Individualism, the ideal of Western culture propagated by the globalization process does not correspond to the ideal view of life according to this universal law of nature.

The historical cycle according to Indian philosophy

Following the basic philosophical premise that the universe is subject to the action of identical universal laws, *Vedantic* philosophy according to Vivekananda (Vol. IV, 1946) says that society develops cyclically. With each cycle society rises to higher and still higher stages and is perfected. Vivekananda divided the whole of Indian society into two classes, the rich (the upper class) and the poor (the lower class). The lower class, the *sudras* or the workers, are the people, the masses; the future is theirs: 'The only hope of India was from the masses, for the upper classes were physically and morally dead' (Vivekananda, 1946b).

Developing on the historical condition, any of the four castes (priests, warriors, merchants and workers) may stand at the head of the state. All persons in each caste possess a combination of three basic qualities of human nature: *satta* (aesthetic), *raja* (strength) and *tamas* (idleness). A specific combination of these three qualities determines the nature of the historical condition. In conformity to the law of nature, the four castes in every society, one after another in succession, govern the world.

Among every ancient civilization, in China, Babylon, Egypt, India or Persia, the supreme power of the society in the first period of their history was in the hands of the priest class. These countries showed features of exclusiveness. The priest class had all the privileges, and based their rule on intellectual rather than physical power. During this period the fundamentals of science were laid and culture was predominantly of an intellectual and literary character. In the course of time the priesthood became obsolete. It outlived its usefulness and became incapable of governing society. According to Vivekananda, 'at a certain time every society attains its manhood, when a strong conflict ensues between the ruling power and the common people' (Vol. IV, 1946). The life of society, its expansion and civilization, depend on who will emerge victorious. The victory of the new progressive forces revolutionizes society.

The rule of the warriors, which replaced the rule of the priest class, was tyrannical, but the arts and social culture reached their peak at that time. The rule of the merchants, which came later, had the advantage that travelling everywhere they spread the knowledge accumulated during the rule of the priests and the warriors. The result of the supremacy of the merchants is the accumulation of wealth. The merchants attained great success in the Western countries. It is due to them (that is, the multinational companies) that the West (Japan as well) became economically strong and able to dominate the world. Vivekananda has described this in a dramatic way: 'the conquest of India by England . . .

that England, whose war-flag is the factory chimney, whose troops are the merchantmen, whose battlefields are the market places of the world' (*ibid.*).

About a hundred years after the days of Vivekananda the picture is the same. However, according to him the supremacy of the merchant class is coming to an end. In future, the supremacy of the workers must emerge. Under it, 'just distribution of material values will be achieved, equality of the rights of all members of society to ownership of property established and caste differences obliterated' (*ibid.*). Vivekananda did not specify how this could be achieved, but mentioned the nature of that ideal state in future:

> If it is possible to form a state in which the knowledge of the priest-period, the culture of the military, the distributive spirit of the commercial and the ideal of equality of the last [period of the workers] can all be kept intact, minus their evils, it will be an ideal state.
>
> (Vivekananda, 1946b)

The first three periods, according to Vivekananda, have already occurred for the world and now the time has come for the fourth.

Philosopher Arabindo Ghosh who has devoted his life to understand and explain the Indian view of life has explained another dimension of the ideal state. The principal contradiction of human life is that between the individual and society or aggregate. 'The essence of the ideal law of social development demand that individual should harmonize his life with the life of the social aggregate' (Ghosh, 1971). The law for humanity is to pursue its upward evolution towards the findings and expression of the 'Divine' in mankind.

In order to achieve this ideal state, one must, according to Vivekananda, understand the causes of the downfall of the colonized world. The causes are perversion of religion, tyranny towards the masses, absence of due education and instructions, underestimating the role of women, and physical and spiritual weakness and inertia (Vivekananda, 1946b). Down the centuries, the rulers and the dominant castes neglected the interests and the lot of the simple people, and that was one of the greatest social evils. Without support of the lower class, there should be no question of serious reforms. Highly developed production and material well-being cannot by themselves make men happy if their 'spiritual civilization' is low. In the West, men are only capable of seeing the external aspects of things according to Vivekananda (Vol. II, 1946). This is the view of history according to Indian philosophy, which expects that in future there will be a socialistic state of the *sudras* (the workers); 'Everything goes to show that socialism ... call it what you will, is coming on the boards' (Vivekananda, 1946b). Batra (1978) has explained why the current era of merchants, or capitalism, cannot have any future; the reason is that the acquisitive tendencies of the merchant class create contradictions in the system by stimulating popular discontent.

In capitalism wealth is being concentrated in the hands of the few. The dominion of the capitalist class today is justified in the name of economic growth and production efficiency. But the resultant deprivations are visible even in developed countries. In the United States, about 12 million people are homeless, one-third of the people cannot afford even primary healthcare, 20 per cent of the children are living below the poverty line, about 23 per cent of the people are illiterate with no security of either job or of life.

Thus, capitalism has so far failed to maximize social welfare through the maximization of individuals' profits. The resultant discontent will grow substantially due to the globalization process, which will intensify deprivation in the pursuit of efficiency across the globe. 'History tells us that when leaders turn ultra-acquisitive and insensitive to the plight of the poor, economic conditions become so bad that revolution is the ultimate result' (Batra, 1978). 'Changes occur in society because of contradictions in prevailing ideology, in its social, economic and political order. These contradictions arise from hostilities between the social classes' (Marx, 1977). Globalization can only intensify these contradictions, bringing the ultimate collapse of capitalism. Indian philosophy does not specify how this destruction of the capitalist system will be achieved. It will be through '... education, spread of knowledge' (Vivekananda, 1946b). Whether the historical cycle, as propagated by Vivekananda will be fulfilled or not, is a matter of debate. Vivekananda has proposed his theory not only for India but for the world as a whole. The increasing hostility of the people across the world towards the 'globalization process' may intensify the cyclical movement of the world history.

Concluding comments

The analysis of Indian philosophy as given in the *Vedanta* doctrine shows that the future cannot be determined by technology or economics. It will be determined by universal laws of nature. If the economic system imposed from outside does not correspond to the national culture or the philosophy of life, it will collapse sooner or later due to its own inherent contradictions. According to the Indian philosophy of life, we are in the era of merchants, or the capitalistic system. This system cannot last forever due to the tyranny, oppression and degeneration it creates just like other systems that came before it.

Relationship between culture or the philosophy of life of a country and its economic and social system is important for all nations. Humanistic aspects of any national culture – that is renunciation, selfless work, sacrifice, work without any attachment to results – does not correspond to the acquisitive consumerism glorified by capitalism which is the philosophy of the globalization process. The essential characteristics of national cultures can be traced to these basic human values signified by the Indian philosophy of life which suggests that the present acquisitive consumerism or the capitalist system controlled by the merchant class cannot last, but will be replaced by an alternative system.

The globalization process in developing countries will bring up some of the nightmares of the last century. India during the British colonial period has demonstrated that already. Due to free trade, India's manufacturing industries during the eighteenth and nineteenth centuries were destroyed. New industries could not flourish. Agriculture in India during the nineteenth century was ruined due to the pressure of imports and fiscal burdens imposed by the colonial government. Famine used to be the order of the day. The situation was the same in other parts of the colonized world.

There are already signs of decline in both agriculture and industry in the developing world due to globalization. Oppressions during the nineteenth century gave rise to the freedom struggles in the colonized world during the twentieth century. Oppressive forces of globalization will create similar popular revolts in the emerging market economies. The popular demonstrations against globalization all over the world signify this.

An economic system cannot work in isolation. It must be supported by the culture of the country. If the philosophy of life in a country does not correspond to the philosophy of the economic system, that system will fail. It is essential to look at the future of India and the world economy from this aspect.

References

Barlett, C.A. and Ghoshal, S. (1989) *Managing across Borders: The Transnational Solution*. Boston, Mass.: Harvard Business School Press.

Basu, D. and Miroshnik, V. (2000) *Japanese Multinational Companies: Management and Investment Strategies*. Oxford: Elsevier.

Batra, R. (1978) *The Downfall of Capitalism and Communism*. London: Palgrave Macmillan.

Bentham, J. (1983) 'Theorie des peines et des recompenses, ouvrage extrait des manuscrits de M.Jeremie Bentham', in *The Collected Works of Jeremy Bentham*. Oxford: Oxford University Press.

Bhagwad Gita (1983) translated by Juan Mascaro. New York: Viking Press.

Brodov, V. (1964) *Indian Philosophy in Modern Times*. Moscow: Progress Publishers.

Bukharin, N. (1917) *Imperialism and World Economy*. London: Merlin Press.

Chatterjee, B.C. (1986) *Complete Works*, Vol. 4. Calcutta: Tuli-Kalam Publishers.

Child, J. (1981) 'Culture, Contingency and Capitalism at the cross-National Study Organizations', in L.L. Cummings and B.M. Shaw (eds), *Research in Organizational Behaviour*, Vol. 3. Greenwich, Connecticutt: JAI Press.

Cardosso, F.H. and Faletto, E. (1979) *Dependencia y Desarrollo en Americana Latina*. Santiago: Siglo Veintiuno Editors.

Evans, P.B. (1991) *Dependent Development: The Alliance of Multinational, State and Local Capital in Brazil*. Princeton, New Jersey: Princeton University Press.

Ghosh, Arabindo (1947) *The Life Divine*, Vol. 1. Calcutta: Arya Publishing House.

Ghosh, Arabindo (1971) 'The Human Cycle', in *Selected Works*, Vol. 15. Pondicherry: All India Press.

Hayes, R.H. (1981) 'Why Japanese Factories Work', *Harvard Business Review*, July–August.

Hayashi, S. (1988) *Culture and Management in Japan*. Tokyo: Tokyo University Press.

Hobson, J.A. (1902) *Imperialism*: A Study. London: Unwin-Hyman.

Hofstede, G. (1993) *Cultures and Organizations*. New York: McGraw Hill.

Hunger in America 2001 Report, (2001) Second Harvest, Chicago.

Kluckholm, C. and Strodbeck, F.L. (1961) *Variations in Value Orientation*. New York: Row Peterson & Co.

Kroeber, A.L. and Kluckholm, C. (1952) *Culture: A Critical Review of Concepts and Definitions*. Cambridge: Harvard University Press.

Kudrle, R.T. (1991) 'The Seven Faces of the Multinational Corporation: Political Reaction and Policy Response', in K.W. Stiles and A. Tsuneo (eds), *International Political Economy*. New York: HarperCollins.

Lenin.V.I. (1917) *Imperialism, the Highest stage of Capitalism*. Moscow: Progress Publishers.

Marx, K. (1977) *A Contribution to the Critique of Political Economy*. Moscow: Progress Publishers.

Miroshnik, V. (1997) 'Diversity in International Managements of Multinational Corporations: The Role of Culture', *Keiei To Keizai*, Vol. 76(4).

Mill, J. (1986) *Elements of Political Economy*. Cambridge: Cambridge University Press.

Mill, J.S. (1999) *Principles of Political Economy*. Oxford: Oxford University Press.

Morita, A. (1992) 'A Moment for Japanese Management', *Japan-Echo*, Vol. 19(2).

OECD, (2001) *Child WellBeing, Child Poverty and Child Policy in Modern Nations*. Paris: OECD.

Report of the U.S. National Center for Children in Poverty, (2001) Washington, DC.

Rolland, R. (1944) *La Vie de Vivekananda el L'Evangile Universel*, Part 1. Calcutta: Vedanta Press.

Russell, B. (1946) *History of Western Philosophy*. London: Routledge.

Smith, A. (1998) *An Inquiry into the Nature and Causes of the Wealth of Nations*, 2 Vols. Oxford: Oxford University Press.

Sunkel, O. (1972) 'Big Business and Dependencia', *Foreign Affairs*, Vol. 50, April, pp. 517–31.
Vivekananda (1946) *Complete Works*, Vols I to VIII. Calcutta: Vedanta Publishers.
Vivekananda (1946b) *On India and Her Problems*. Calcutta: Advaitya Ashrama.
Weber, M. (1946) *The Religion of India*. Glencoe: The Free Press.
Weber, M. (1993) *Protestant Ethics and the Spirit of Capitalism*. London: Routledge.

14

Psycho-Pathology of Globalization and the Indian Surrender

S.K. Chakraborty

Caveat emptor

Here are some sacred gospels in marketing: customer focus, customer delight, customer is king; or some other variant of like nature. Yet there is also this warning: *caveat emptor*, buyer beware! Of course no one turns back to warn: 'seller be honest'. What a jumble of contradictions!

Globalization as a panacea has been marketed vigorously over the whole decade of the 1990s. Has it given me, a consumer-citizen of India, delight? Made me feel I am a king? Or, has it duped me into avoidable, dissipative material slavery? Has globalization made me, an entrepreneur-citizen of India, feel secure and optimistic? Or, am I on the verge of being wiped out of existence? So, let me enumerate a few direct consequences of globalization upon the common citizen of a country like India:

- Sony, Akai, Samsung etc. have completely eliminated domestic Onida, Weston, Sonodyne etc. in the TV market;
- Ambassador, Fiat etc. have been squeezed into a tiny corner by Matiz, Cielo etc.;
- Cereals like puffed or flattened rice are being replaced by Kellogg flakes; Thums Up and Gold Spot, by Pepsi and Coke;
- Small-scale industrial units are crumbling like a house of cards;
- The cult of efficiency is being invoked blindly, and large-scale industrial units are downsizing manpower pretty ruthlessly;
- A compensating increase of employment opportunities in the real economy is being ruled out;
- Thefts, robberies, juvenile crimes and adult suicides are becoming frighteningly frequent in one's neighbourhood or circle of acquaintances;

- Regular Miss Universe/Miss World title winners from India in recent years – a striking symptom of globalized marketing by any means;
- Consuming by borrowing, by flaunting the credit card and the rest are gradually becoming the in-thing;
- Most TV serials and commercials are engaged in transmitting vulgarism and depravity to tender minds;
- A rapid rise of disruptive, wayward individualism in the name of personality development;
- Careerism and money-lust are fast becoming the only keynotes of education;
- Competitive aggression in interpersonal and intercorporate dealings is reaching alarming levels in schools and upwards;
- Disquieting rates of imitative marital infidelity, divorces and the like in Indian homes; and
- Unimaginable computer-related academic crimes in institutions of higher education.

So much then for the phoney gospel: 'customer is king'! While there is little hope for the 'seller be honest' ideal to become actionable; at least the buried 'buyer beware' maxim must be exhumed and adopted by all developing countries. And this they can do only through stern self-discipline at the highest political and bureaucratic levels. Let us examine the matter more intensely.

The psycho-pathology of tears-for-the poor

A central point should be clear right at the start: the cry 'globalize or perish' was not raised by developing, poor nations. It is the rich, developed economies of the world who have orchestrated the blitzkreig of globalization. The immediate next question is: what inspired them to mesmerize the rest of the world with this *mantra*? The answer is: to gain markets in high-population large countries for their MNCs/MNEs to enable them to earn higher rates of return on their assets and maintain the rising curve of consumption standards in their home countries.

Fundamentally, this phenomenon reflects the ever-sophisticated ploys stemming from the possessive vitalistic psychology of those who already have much. They are mortally fearful of not being able to keep up the furious pace once set by them, and they are also irreversibly addicted to it. It is this pathology of fear plus addiction which truly explains events like the Falklands War, the Gulf War and so on.

Another symptom of pathological proportions linked with globalization is the strange infatuation of key players with abstract wealth, with money flows which have no links with real wealth. Today a few people are becoming millionaries/billionaires almost overnight through merely speculative market valuations of their shareholdings. Ephemerality is of course the natural correlate of such wealthiness. Thus, the Nasdaq index had peaked at 5000 around March 2000, and a year later it has declined by 60 per cent. It may fall even more.[1] These global financial markets remain insulated from the influence of national governments in respect of local imperatives. Consequently, grassroots developmental needs fail to be addressed. Rich-poor divide goes on widening, and global poverty increasing – as highlighted by successive UNDP Development Reports.[2] Vasco Da Gama from Portugal had landed in Calicut, India, in 1498; this was not a soft-landing of goodwill and culture, rather, more than 500 years ago this event had heralded a process of bloody and open exploitation and domination. With modern techno-economic globalization the very same psycho-pathology of 'might is right' is showing up shrouded in complex guile and subtle sophistry. Here is a relevant example.

A recent *Economist* editorial scales the summits of hypocrisy and arrogance when it dubs the Seattle, Washington DC and Prague protesters against the 'mighty forces of globalization' as a bunch of 'great nuisance makers'. With supreme condescension it avers that the activities of the anti-globalization groups foreshadow 'an unparalleled catastrophe for the planet's most desperate people'. Defending globalization as a moral crusade, the same editorial ends up with copious tears for the world's poor![3] The UNDP *Development Reports* – are you listening?

A somewhat more balanced and elaborate discussion on anti-capitalist protests, compared to the vituperative editorial, follows in the same issue of the magazine. Tears for the poor are of course not wanting in this commentary as well. But it at least concedes that the protesters do reflect 'popular concern about the hard edges of globalization': fears about leaving the poor behind, harming the environment, priority to profits over people, marketing of suspect genetically modified foods and the like. At the same time the piece laments the concessions the Bretton Woods institutions and the UNDP are already making to the intellectually incoherent (!) noises made by an unaccountable motley of NGOs in their revised policy orientations. As if economic globalization itself is a masterpiece of transparent clarity and candour! In the fiery words of the defenders of globalization themselves it is a mighty engine which must be fuelled by untrammeled 'individual liberty'. We may read into this the arrogant demand to crush everything that comes its way as it hews its tortuous, clandestine path through sustainable societies and cultures across the globe. What a bitter travesty of the sweet phrase 'global village'!

I had a symbolic personal experience of this pathology of go-as-you-like globalization at an International Conference in Stockholm on 1 June 2000. After one of the high-priests of IT, which is claimed by the inveterate globalizers as the main (and add, blind) driver of the juggernaut of globalization, had finished his dazzling presentation, the moderator of the session asked the audience if, at the end of it all, they felt (a) happier, (b) unhappier, (c) just about the same compared to say 15 years ago. For (a), in a 900 audience, perhaps a 100 or so hands went up, for (b) only 2 went up. I was one of these two. The moderator asked me: 'Why do you feel so?' I replied politely: 'Because I am feeling increasingly distracted and disturbed in my living'. The presenter got angry, and began to walk up to me. I was startled. Reaching me he asked, 'why do you not leave this world?' 'Where do I go?', I asked in return. 'Perhaps to the Himalayas' he advised. I kept quiet. The audience remained a mute witness to this brief drama. In the same vein, newspapers in India had reported (sometime in 2000) the ambassador of the world's richest economy as making an assertion to the effect that: 'What is good for . . . must be good for the world'. This then is the sinister psychological face hidden behind the mask of tears-for-the-poor worn by the mainstream protagonists of globalization.

Triune guna dynamic and globalization

Just a moment ago I used the word 'blind' besides the earlier ones of 'fear', 'greed', 'addiction' and 'arrogance', to characterize the psycho-pathology of economic globalization. This whole recipe merits a brief, yet deeper probe. One of the central theories of human psychology in India is that of the three *gunas* (*tri guna tattwa* in Sanskrit). Rendered into English this means the theory of 'triune psychological energy-forces'. It had been formulated by the seer-psychologists of classical India after prolonged analytical observation of human behaviour, and confirmed simultaneously by thorough experiential realization. This triune energy-force theory represents a luminous peak in the vast terrain of holistic spiritual psychology in Indian culture.

The first of this triad of psychological energy-forces is called *sattwa guna*, the second *rajas guna* and the third *tamas guna*. *Sattwa* is constituted by luminous understanding, inward depth and serenity, outward steadiness and sincerity. *Rajas* is of the stuff of blind activity, inward shallowness and turbulence, outward grabbing and deviousness. *Tamas* is made up of darkness and inertia – the opposite of both *sattwa* and *rajas*. Each personality profile is always an integral composite of these categories of three *gunas*, but their relative proportions always vary from one entity to another. This variation explains the salient differentials in the manifest behaviour and unmanifest dispositions amongst children of the same parents, members of the same group, citizens of the same country, and amongst countries of the same world.

Dominated by *sattwa guna* in the triune blend, the revealed profile would tend to be patient, far-sighted, contented, caring much more for inner/spiritual progress. Perfection would receive priority over success by instinct. Governed by preponderant *rajas guna*, the personality-in-action would tend to be mercurial, short-sighted, perpetually wanting, lured much more by ephemeral external victories at the cost of defeats within. Success without, by clever manipulation, replaces perfection within as the goal of living. Under the influence of a dominant *tamas guna*, however, action-orientation is in short supply. If *rajas guna* be visualized as a hyperactive misguided missile, *tamas* may be seen as a defused missile, incapable of detonation.

With the help of this theory of triune *guna* dynamics we are able to draw faithful portraits of national, institutional, familial as well as personal character and ethos. Historically speaking, India appears to have served as a good melting pot to study and interpret both national and personal personality profiles in terms of *guna* theory. Thus, from Alexander to Vasco Da Gama, from Tamer Lane and Nadir Shah to Dupleix and Clive – for over 2000 years India has been the object of assault by *rajasic* cultures from the West. Three centuries ago almost all the West European countries were found fighting among themselves on Indian soil for her spoils. What a bizarre spectacle! On the other hand, from Ashoka to Ashwaghosha, from Ram Mohum Roy and Swami Vivekananda to Rabindranath Tagore and Mahatma Gandhi, India had forayed into the East and the West with the gospels of culture and peace. All this was a saga of *sattwic* enrichment by India of other countries. No guns and swords were employed in this transborder intercourse. This then is the learning point: the world has a choice between the *sattwic* globalization of humane intercourse demonstrated historically by India, and the *rajasic* gobblization now perpetrated by the rich economies in one form or other throughout the last two millennia and more.

We can glimpse a few thin shafts of light pointing towards the possible dawn of an era of *sattwic* globalization of a humane type, not just the economic brand. For, in the latter selfishness is always the core. So it must end up in beggar-thy-neighbour *rajasic* globalization. That some of the governments in highly developed economies, and some of the global institutions, despite the barrage of strictures from the *rajasic* defenders of gobblization, are now beginning to recognize the voice of dissenters is one such ray of hope. The slow but steady awakening of the need for maintaining the distinctive cultural identities, which have provided meaning and rootedness to each entity in the endless ensemble of community life processes on earth, is the other streak of light.[4] The third hopeful sign is the emergence of a steadily growing group of researchers and writers in the developed countries themselves who are daring to raise *sattwic* questions against their own *rajasic* peddlers of the magic wand of gobblization.[5] In this respect the irresponsibility of mainstream scholars and advisers in countries like India is absolutely shameful. Neither *caveat emptor* nor seller be honest makes any sense to them.

It is also important to carefully note that the technology-driven, high-speed, new economy model being followed by the developed economies is proving to be a psychological blight for the people of such countries themselves. The theory of *guna* dynamics does indeed foretell this outcome of blind and greedy *rajasic* dynamism, without the serene guiding light of *sattwa guna*. Vanishing leisure, proliferating lifestyle diseases, stress and meaninglessness – all these deeper illnesses are the dominant afflictions of today in these societies. With the displacement of the wisdom model of 'simple living, high thinking' by the smart model of 'complex living, low thinking', we can appreciate the following cry of near-despair:

> Yet these gains [out of hi-tech] have come at considerable cost... the hyperaccelerated world in which we shockingly now find ourselves... we seem to have less time with every passing year. We now work seven days per week, answer e-mails on Sunday morning, take our beepers and cellphones with us during the few hours we have left to play. And there is that gnawing fear that it is going to get much worse before it gets any better.[6]

Is it sensible and wise for the intrinsically *sattwic* India to allow herself to be swallowed up in the quicksand of such insanity while running the unwinnable *rajasic* economic race where the shots are called by those very countries which are bemoaning their own psychological plight?

Sri Aurobindo has offered us impeccable insights into this essentially *rajasic* psychology acting as the driving force behind economic globalization:

> Its [the *rajasic* mode of Nature] fruit is the lust of action, but also grief, pain, all kinds of suffering; for it has no right possession of its object . . . and even its pleasure of acquired possession is troubled and unstable because it has not clear knowledge and does not know how to possess, nor can it find the secret of accord and right enjoyment.[7]

> . . . if the *rajasic* qualities are given the upper hand, cultured to the diminution of *sattwa*, then the trend of *karma* and its results necessarily culminate . . . in the highest exaggeration of the perversities of lower nature.[8]

> The man, if he does not stop short and abandon his way of error . . . he can no more reverse the fatal speed of his course because of the very immensity of the misused divine power in him.[9]

Subjecting our experience of the ongoing fury of economic globalization to the X-rays of Sri Aurobindo's telling exposition of *rajasic guna* dynamics, we are able to penetrate right into the heart of the pathology of globalization which is turning it into gobblization. Unless courage is shown to pick up this critical psychological clue, and to develop a vigorous programme for leadership of all types all around to learn self-management on the basis of this theory, a lot else that is being espoused, though often well-intentioned, will remain platitudinous and ineffectual. If we continue to be weak, defensive and apologetic, we might be paving our road to hell with good intentions.

Globalization, happiness and India

Viewed from the perspective of a low need–low greed, longlife–low profile Yogic culture like India's, the hurricane of globalization during the last two decades of the twentieth century may be seen as flooding the world with rising materialistic wants. Of course the rhetoric is all sweetened with surging sympathy for the poor, the third world and so forth.

An important conceptual clarity is in order here: the development agenda proclaimed by globalization persistently mistakes 'standard of consumption' (SC) for 'standard of living' (SL). A 1999 World Happiness

Survey placed Bangladesh on top in terms of happiness, the USA 46th in rank, and India occupied 5th position. It is well-known that SC in Bangladesh is among the lowest in the world. But this has not robbed them of happiness. Equally well-known is the fact that US citizens enjoy the highest SC. Yet they score so low in terms of happiness.[10] Such data yield the plausible hypothesis that SL in Bangladesh and India still preserves, somehow, such non-measurable qualities of life as make for greater happiness than that in high SC countries. The real danger is that globalization of greed, pursued by high SC–low SL nations in their own apparent interests is soon going to destroy the happiness in high SL–low SC countries.

The Internet summary of this Happiness Survey goes on to report that people in most rich countries including Japan, the Netherlands, Switzerland, Canada and Austria are much more unhappy than their counterparts in poorer countries. The British, although having twice as much money to spend in real terms, perceive that their quality of life has not improved. This has belied the nation's earlier belief that money could bring happiness. In fact Britons are today generally less happy than they were 10 years ago. The seductive economic siren is making them unhappier because of emotional penury caused by consumerism and breakdown of the family.[11] These are exactly the plagues which are now breaking out in so-called poor countries like India.

Therefore the predatory globalization that is now sweeping across the world must answer at least two tests: 'material sustainability' (MS) and 'psychological sustainability' (PS). Large masses of empirical data from bodies like the World Watch Institute and several other research studies converge on the conclusion that technology-driven, consumption-inspired economic development, pushing ruthlessly beyond national borders for markets, is materially unsustainable. After all information itself cannot feed and clothe and shelter people. What it can do is to whip up unfelt, unimagined wants at great speed in the remotest corners of the earth. The explosion of global material greed, beyond legitimate basic needs, may serve some ill-conceived corporate and national interests of a clever few for a short while. But the long-term negative impact of such short-sighted strategies on the earth's MS is already alarming. Recall the 11 million square miles of hole in the ozone umbrella over the Antarctica as reported by both Nasa and the World Meteorological Organization in September 2000.

There is another tragic dimension of this assault on MS – the extinction of local cottage and small-scale entrepreneurship. As mentioned earlier, India is today witnessing a rapidly swelling number of dying or dead backyard enterprises in every conceivable product – zip fasteners, razor blades, cereals, bottling plants for aerated drinks, fountain pens, televisions, earthen cups for tea/coffee etc. Let us not speak of cars and computers. The result is an inexorable shrinking of the decentralized, mass-employment, grassroots economic base of the country. The apparently benevolent activity of rural marketing is really a well-planned thrust by powerful MNCs to make the simple-living rural people greedy for non-essential consumer goods. Their motto is: 'we need your greed for our greed'. All in all, the low-profile, close-to-soil, environmentally compatible units of economic activity are fast disappearing. This is certainly lethal for the intrinsic MS of countries like India.

A practical implication of the above reasoning is that globalization of business and finance, in order to preserve MS, should not go beyond basic infrastructural projects in developing countries. They would chiefly be power, roads and rails, irrigation and the like. The whole range of consumer goods, durable or otherwise, must be left free from inroads by MNCs. One of the ways local economies, from Scotland to Russia, are now exploring is going back to some sort of barter system to insulate themselves from such invasions.

As for PS, the *rajasic* champions of globalization (gobblization) seem to be oblivious of an essential law of being: 'man does not live by bread alone'. This perennial psychological truth had been translated into daily living by sustainable cultures like India and others with vigilant creativity. But the wanton creativity of the present era is substituting for this principle, with bullish fury, the false mantra: 'man lives by goods (and information) alone'. The more the human being is trapped in a ceaseless chase for creature comforts without, the greater is the estrangement he/she suffers from intrinsic wealth within. The grinding centrifugality of 24-hour modern living, caught in a dizzying spin, makes us destitute of a centripetal anchor. This constitutes an existential disempowerment of the deepest kind. The pathology of external self-indulgence that the 'globalization-market economy' cult relies exclusively upon, robs us of our internal SELF-sufficiency (e.g. USA *vs* Bangladesh). The motive of self-indulgence, for that is what really underpins globalization for the sake of business profits, national GNP, and executive stock options, is both materially and psychologically unsustainable. In the long run, therefore, the strategy of 'frugal external consumption-cum-ample inner living' alone is both materially and psychologically sustainable. In other words, higher SC will induce both lower MS and PS. Happiness comes more from *sattwic* 'giving', not from *rajasic* 'grabbing'. This implies lower SC which will spell itself into both higher MS and PS.

Here are some findings sampled from a recent field study by an Indian NGO called 'Society for Integrated Development of Himalayas' (SIDH) which show how the plagues of globalization are eating into our society from beneath:

1 'The literate person is selfish and greedy and wants good things... He/she aspires for consumer items and imitates the West. The literate person is more prone to individualism. The literate people have a greater tendency to drink, smoke, gamble, see films and generally spend a lot of money on entertainment.'[12]
2 'The literate person is less responsible and lacks discipline whereas the illiterate persons are more responsible towards elders and family. The literate have less respect for the elderly, answer rudely, do what pleases them, order people older than them, use more abusive language.'[13]
3 'Children developing a strong attraction towards expensive consumer items (like Nike, Reebok shoes) which parents find difficult to afford.'[14]
4 '...the urban parents lamented that their children have become spoilt (*bigad gaye hain*), the rural parents expressed their despair in much stronger terms that their children have been completely ruined (*barbaad*) by education.'[15]

In another country-wide study of the impact of globalization on India by a British researcher we find observations like this:

1 'To cite some of the current status signifiers (none of them a necessity), a pair of Nike trainers costs about Rs. 3,000; a box of Kellogg's Com-flakes is roughly Rs. 100 and a pizza in Pizza Express is about Rs. 300...'[16]
2 'Especially amongst the ambitious young, the obsession with international brands, contemplation of exorbitant prices and what is taken for international culture has become almost pathological.'[17]
3 'The reason for increasing fascination for overseas education is that what is both desired and derived is an exposure to globalized commodity culture, rather than the culture of organic communities in Western settings.'[18]

The above excerpts show that globalization of the kind now engulfing the world is destabilizing and unsustainable from all points of view – material, ecological, social, psychological and spiritual. At the same time, it is this very pervasive subversiveness of greed-based globalization which is also causing a new awakening to the need for spirituality. Spirituality has to cast its sobering light on the red heat of money lust. Of course spirituality itself is often getting commoditized, for example some authors are already talking about 'spiritual capital'. Having quickly discovered that knowledge or intellectual or emotional capitals are not delivering whatever they promised, some of us are trying to extract the same promises from 'spiritual capital'. This tendency has to be corrected before it becomes another tragedy. For, if spirituality also is enslaved to material greed, no saving alternative will be left for the world.

The murky backyard of globalization

A visiting fellow from a New Zealand University at our Management Centre for Human Values, who has been teaching business ethics there, had recently narrated to me the inside story of how the developed nations had resorted to undemocratic manoeuvres ensuring that the CEO of the WTO would not be appointed from a developing country. In fact the post went to a person from his own country – New Zealand. If the WTO were such an innocuous, toothless body as Rugman asserts, why should such inside manipulations have taken place at all? He says about the WTO: '...in reality it is an understaffed and overworked technical bureaucracy with no political clout at all'.[19] I recall, too, the remark made to me by a Professor of the Copenhagen Business School (in 1997) that the WTO is one of the most vicious things to have happened so far as the developing economies are concerned.

Quite recently a British commentator has been more brutally frank than most regarding the WTO. As a part of the WTO, the General Agreement on Trade in Services (GATS) was signed in 1994, covering 160 sectors – virtually the entire world economy. The first stage of GATS requires governments to open up their service sectors (for example opening up the banking and insurance sectors in India). Developing countries face unfair negotiations, complex agreements, and no assessments on the after-effects of such opening up. The second aspect relates to regulations for large MNCs operating in these sectors in developing countries. Here, as the author points out, the commitments for continued greater liberalization are going to strip the rights of democratic governments for stronger regulation or greater public sector involvement in their national interests.[20] Thus, from 1 April 2001, pressurised by the WTO, India has been forced to permit imports of 715 items without any quantitative limits. As a sop, exports are likewise also to be made unrestricted. The whole strategy is going to turn India again into a finished-goods-importing and a raw-materials-exporting economy. This is harking back to colonial days.

A report published in 1999 by *Crossing Boundaries* revealed some staggering inside information and inferences drawn from the secret activities of powerful, globalizing MNCs. It highlighted the sinister designs of the 29 rich countries comprising the OECD in formulating the Multilateral Agreement on Investment (MAI). They were continuing to hold these secret negotiations since 1995 until in January 1997 its provisions leaked out on the Internet. The MNCs of these countries had been backing up the MAI wholeheartedly. It is only the concerted and dedicated efforts of some NGOs which spilled the beans. The report sums up the gist of the MAI as follows:[21]

> The MAI is, in its most basic form a 'bill of rights' for transnational corporations. It would create strongly enforced, legally binding rules limiting the right and ability of sovereign governments to regulate foreign investors and corporations.

If the MAI were to become operative then, among many things as the report notes, be it noted by India, that 'local businesses everywhere, *whatever their size*, would be forced to compete with transnational giants on an equal footing'; and also 'local, state, provincial and national governments could be sued by corporations – compelled to pay "damages" and to overturn laws that protect their citizens from corporate exploitation' (my emphasis).

It is pertinent to ask here: what role is played by ethical conscience-keepers like Transparency International to reveal, measure and rank-order the closely guarded machinations of the rich nations who conspire to evolve MAI-type and GATS schemes? Without denying that there is a measure of truth in the corruption rank-orderings produced by such bodies where countries like India appear in a very bad light, it must be understood that it is mostly the executives of the MNCs of these very MAI GATS economies operating in developing countries who are the basic source of opinions polled in these surveys. Pots calling kettles black! And, to wit, it is often these MNCs themselves which mastermind the mega corrupt deals through the backdoor to gain access into new markets. In September 2000, during a dinner by my host in Geneva, a former faculty member in a French business school and a director in a large Swiss pharmaceutical company, described most such MNCs as 'sharks and crooks'. He has resigned from the Board in disgust, and now devotes himself to real grassroots work in the field of healthcare for the poor in India and elsewhere.

To shield this dark face of the MAI or WTO conclave, analysts like Rugman lay the blame for scuttling its implementation on the 'negative influence of non-governmental organizations'.[22] For people with his perspective, MAI or GATS-type international accords are required to replace individual country regulations which 'distort trade and investment patterns'.[23] 'Distortion' from whose point of interests, may we ask? For them state power (of a given country) discriminates against MNCs across many sectors – Rugman himself provides this answer.[24] For such audacious protagonists of MAI and so forth in G7 countries of the world the 'shallow integration' achieved by free trade must be replaced by 'deep integration' through free investments as well.[25] Thus the richest 7 or 29 economies of the world must have untrammelled sway over the globe to further their own interests.

Several decades ago, Ananda Coomaraswamy, the famed art-critic of India settled in Boston, had critiqued industrialism and trade with penetrating, far-reaching insight. 'Inspired tradition' (we call it *sattwic* tradition), Coomaraswamy had averred, 'rejects ambition, competition and quantitative standards'. 'Modern civilization' (we term it *rajasic* civilization), on the other hand, is based on social success, free enterprise (devil take the hindmost) and quantitative measures. The first one meets man's needs which are 'but little here below', whereas the second one panders to unlimited wants multiplied artificially by advertisement. For profits, ever-expanding markets beyond national borders are needed to sell the surplus. Historically, such globalization of greed has destroyed many civilizations – the Spaniards did so in South America, the Japanese in Korea and by 'white shadows in the South Seas.'[26] Coomaraswamy goes on to pose the supreme ethical issue with passionate intensity:[27]

> Would it not be better if, instead of tinkering with the inevitable consequences of world trade, we considered its cause, and set about to re-form our own civilization? Or shall the uncivilized forever pretend to civilizing missions?

We might as well, in the present context, just substitute the current words or phrases 'globalization', 'unglobalized' and 'poverty-reduction missions' for 'civilization', 'uncivilized' and 'civilizing missions'. Then Coomaraswamy's ethical question gets adequately recontextualized for what is happening now.

At an international symposium on Applied Ethics held at the Management Centre for Human Values (MCHV) in Indian Institute of

Management, Calcutta (IIMC) in February 1998, David Kimber presented a wide spectrum of ethical issues in international technology transfer deals from the North to the South. ITT is of course a major dimension of globalization. During the presentation he narrated two specific examples of ITT from a book by Dyson – one for digging a well somewhere in Africa, and the other for erecting a magnum telescope somewhere in Russia. Both the projects were utter failures serving no useful purpose. Yet, as Kimber concluded, they were pushed through because of dis-values amongst the officials incharge – both among suppliers and buyers. These dis-values were vanity, egoism and selfishness,[28] all springing from unbridled activation of the *rajasic guna* alone. Unethicality is a natural child of such actions.

Elizabeth Gerle has characterized the unethicality of globalization as 'marginalization of the poor and the feminization poverty'. She narrates a case from the Philippines: at an APEC meeting in October 1998, where boastful claims were made about the benign results of liberalization and globalization, especially for women. But the actual picture, in the Philippines for example, has been quite the opposite. By agreement between the WTO and APEC, cultivation of rice and corn has been replaced by that for export items like asparagus and flowers. Also, agricultural land was being converted into golf courses and tourist amenities. Women employed in all such original activities of the country are now becoming caddies and entertainers in clubs. They have been converted into reserve labour with below-minimum wages and no job security. She writes further: 'It is not uncommon for poor women to trade sex for some rice or milk; the commercialization of education is forcing many women students to work as call-girls during the night'.[29] Similar instances were quoted by us with respect to India in an earlier section.

But all this gets submerged in the torrent of frothy rhetoric, clothed in abstract statistics, from the prophets of globalization. Who would tell them that breaking down the holistic psycho-social deep-structures of different cultures by chanting the false slogan of poverty-amelioration is by no means a civilized or a civilizing mission! It is rather arrogant of shallow scholars like Rugman to contemptuously dismiss the critical socio-psychological insights offered by the deeper students of globalization like Giddens.[30] The aggregated, quantitative reports of economists and business strategists completely miss the irreversible but unseen corrosion of the social-psychological foundations of a whole variety of integral and sustainable cultures – all caused by economic-financial globalization of a barbaric character.

At the second annual Forum 2000 Conference in Czechoslovakia (October 1998) the Malaysian minister of special functions had defended the imposition of strict new currency controls by his country. In 1997 the Malaysian currency had sunk alarmingly low after the removal of a host of regulations under the advice of international agencies which had preached the virtues of free capital flows but did not warn about fickle investor sentiments. The minister also commented critically on how many in the West had then celebrated with glee and relish their windfall gains from the collapse of Asian currencies.[31] Any Transparency International indices to rank order such good Samaritans? It is the very essence of *rajasic guna* to derive pleasure from one's gains at the expense of others.

But such normative ways of judging the effects of globalization are foreign to most of its advocates. A recent British text-book[32] on the subject visualizes the chain of current happenings as a 'reconstitution of the world into a single social space'. It is this process which is the cause of economic globalization. And this, in turn, heralds a higher level of intensifying crossborder networks in every aspect than ever before. As a result, larger sections of humanity in both rich and poor countries, more in the latter, are being expelled from the thickening network of human social and economic interaction. Later in the book, chapters 6 and 7 especially, the author seems to overwhelm the reader with copious new jargon – for example deepening phenomena, compression of the world, time/space distantiation, financial deepening, flexible accumulation and so on.[33] Impressive as they are, all of them struggle to express merely the 'is'. There is no engagement with the 'should'. Are all these is's beneficial or harmful for mankind as a whole, in existential terms? Who are the agencies and persons that are instigating these changes? What are their motives? Are the intended beneficiaries being honestly informed and consulted by these changemasters? Or, are these processes being forced on them by hook or by crook? Or they are being duped by sugarcoated reform recipes? Why at all should there be a 'single social space' across the world? What does this mean for cultural sustainability, diversity and freedom? Does not the entire agenda smack of an anti-human-rights movement?

The ontology-epistemology of globalization

A brief, sort of side-glance, at the ontological-epistemological foundations of globalization should help towards a firmer grip on the causes of the pathological outcomes outlined above.

Ontology, as we know, concerns itself with the nature of our being. Mankind creates home and society in such a way as to reflect and enhance that image of himself which he holds closest to his heart. This objectification of his existence is unlikely to be any one pure type; yet the central drift will not be missed. Thus, all ancient and medieval human settlements in India grew up at the confluences or on the banks of big rivers. A massive temple complex would be the nucleus around which every aspect of individual and group life would be organized. Why? Because, the over-arching conception held about the human being was that he/she is, at the highest, deepest, truest plane, a spiritual being. Every component of social existence must subserve the consummation of this *sattwic* end. This ultimate end being never in doubt, trade and commerce, education and medicine, local government and defence – everything was integrated into the hub of human life: to realize the Spirit. Insistence on this view of the being encouraged the scrutiny and judgement of all secular activities against the standard of the sacred, the spiritual. This prioritization forced economic activities (*artha*) to observe limits. Secular desires were legitimized but subjected to strict moderation.

But modern *rajasic* civilization, in its present phase, has exalted the material-economic being in man to the apex. The hub of settlements by urbanized, economic man of today is the tallest bank building in a concrete jungle. The hub in the home is the PC. The eternal has been totally displaced by the ephemeral, the inner by the outer. A thoroughly exteriorized, unstable man – this is the ontology of being which secretly but surely underpins the volatile economic-financial globalization process.

Parallel to this radical slide in the ontological viewpoint from the *sattwic* to the *rajasic* we can discern a huge swing in the epistemological position as well. Epistemology has as its field various methods of knowledge. The post-enlightenment fascination with intellect-reason as the only gateway to knowledge has crystallized today in the kind of globalization we are experiencing. Intellect/reason is precise in conceptualizing within a very small circle. Even here during implementation of the reasoned concept intellect stumbles and makes a mess of the outcome. Both the knowledge acquired and knowledge applied are short-term, fragmented and problem-creating.

In earlier times, when stable and sustainable societies were first founded, the major mode of knowledge was by perception, not conception. Perception is more organic, concrete and wholistic than conception. Perception is independent of reasoning – it is direct and intimate. Intuition is unclouded perception. It is amenable to disciplined cultivation. The chief psychological preparation for holistic intuitive perception is silencing of mental thinking or reasoning, coupled with purification of feelings and emotions. Of course by now this has become an anathema to the hectic, self-indulgent *rajasic* modern man. Yet all great truths of the world have emerged from *sattwic* characters who could silence their shallow, noisy, surface intellect and allow self-existent holistic perception to descend into their purified being. Conventional intellect and reason did work for them, but only at the instrumental level under the guiding light of such holistic perception. To put it simply: it is the holiest men (women) of the world who have been the most holistic as well.

This very short detour towards the ontology-epistemology perspective about globalization might awaken us to both the depth and sweep of the self-created problem confronting the human race. The same shaft of light could well be cast on technology too which is so closely linked with the march of globalization. For us, the call is clearly one about a renewal of priority for the Spirit-being of man, and of a re-enthronement of the trans-rational perceptual process which can make knowledge wise as well. This project for the human race is far more critical for its true evolution than the human genome project and others of its genre.

Conclusion: India's surrender

Why blame the rich globalizers from the North alone? Countries like India must acknowledge a large share of such irresponsible global behaviour. Their political and bureaucratic leaders, out of myopic and blatant self-interest, have been bartering away the holistic inherent wisdom of sustainable cultures for the reductionist time-serving rationality of relatively immature cultures. I could not but humbly agree when a professor at the Notre Dame university sadly confided to me: 'It seems we have lost India as a partner in the crusade against MNCs' (April 2000). Left to themselves, the newly-independent and politically awakened developing nations could have achieved a more egalitarian redistribution of wealth amongst the poor and wealthy sections of their respective populations. But the imposed global economic reforms from the North have been skewing the distribution pattern more sharply in favour of the haves as against the not-haves. It is often discovered that most mega FDI projects (like Enron's power project in Maharashtra, India) are heavily loaded against the receiving country. Ironically, it is the ex-World Bank/ex-IMF top executives from such countries, later occupying high-level policy-making posts in national governments, who play a key role in formulating and approving such exploitative deals, being outwitted by the more sophisticated negotiating skills of

MNCs. A large part of India's predicament is thus of her own making: marrying in haste and repenting at leisure!

The famed economist J.M. Keynes is quoted as having said these words driven, it seems, by some premonition about present-day globalization:[34]

> I sympathize, therefore with those, who would minimize, rather than with those who would maximize, economic entanglement between nations. Ideas, knowledge, art, hospitality, travel – these are the things which should of their nature be international. But let goods be homespun whenever it is reasonably and conveniently possible, and above all let finance be primarily national.

What a miserable localizer must Keynes be, looking in the curved mirror held up by the intrepid globalizing economists of today such as Rugman or Hoogvelt, mentioned earlier.

Long before Keynes, the sage-poet Rabindranath Tagore, who used to be called a 'global', a 'universal' poet in his own lifetime, offered the following critique of modern economic society:[35]

> This is the hideousness of modern commerce: it does not stimulate men into a healthy and normal activity of production, but organizes and makes use of them by mixing and mangling their minds; it is shabbily parsimonious in all that is connected with human life and extravagant in all that tends to the multiplying of market wares.

It may be mentioned that Tagore had travelled all over the globe several times, and the above words were spoken by him in 1920 after his return from China. As if anticipating the much-hyped, upstart rich persons of the world today he uttered this caution of sanity:[36]

> I believe and hope that in the range of all literature and art there is no single instance of a monied man being glorified for the mere sake of his money. Our *Laxmi* is not the goddess of the cash balance in the bank: she is the symbol of that ideal plenitude which is never dissociated from goodness and beauty.

India is now veering away from Lakshmi and avidly courting Kuvera (the Reserve Bank of India had chosen Kuvera in its emblem long ago) – thanks to the blitzkreig of globalization. But does it speak well of our proactive, true freedom?

Five and a half decades after Tagore's address, the renowned cybernetician and system theorist, Stafford Beer, who had served for many years for the British government in India before 1947, was invited to deliver the Zaheer Foundation Lecture in New Delhi (1974). Alluding to his Chilean experience, Beer had queried India with passionate sincerity:[37]

> I was appalled, when I first arrived in Chile, to find the whole middle class . . . already held tight in the grip of consumerism, so that the worth of life was already to be measured in terms of cars, television sets and refrigerators. The devastation wrought in the quality of life in the advanced industrial countries by such use of science and technology is plain to see. *Do developing countries really wish to follow the same road*? (Emphasis added)

Did the policy-makers in New Delhi take to heart this simple yet fundamental question posed by Beer – as late as 1974?

Sri Aurobindo wrote a series of 35 incisive essays on 'The Ideal of Human Unity' during 1915–18 in *Arya*. With his yogic, supramental, holistic intuition he asserted that 'the drive of Nature, the compulsion of circumstances and the present and future need of mankind make it inevitable'[38] – some kind of world-union. Yet his sagacious vision was not one of an artificially manipulated and imposed single social and economic space, but of a single 'psychological unity' which allowed a free association of multiple 'national sentiments and long-established local units'.[39] The greatest obstacle against such a single psycho-spiritual space, as correctly foreseen by him, was 'national and imperial egoism' which would find it extremely hard to forsake the 'instinct to dominate', to abandon the 'advantages of commercial exploitation for self-enrichment'.[40] He also had the courage to point out:

> the claim of Europe, not yet renounced, to hold the rest of the world in the interests of civilization, by which is meant European civilization, and to insist upon its acceptance as a condition for the admission of Asiatic races to any kind of equality or freedom.[41]

This prophecy is being proved true to the hilt today – only the eye of the storm has largely shifted to America.

In another series of writings during the next three years, 1918–21, Sri Aurobindo offered a masterly and penetrating interpretation of Indian culture, unequalled by any other similar work. Addressing the problem of an India caught in the blinding storm noted above, he had again warned us:[42]

> If we take over ... that terrible, monstrous and compelling thing ... European industrialism – unfortunately we are being forced to do it – ... we may under favourable circumstances develop it by our wealth and economic resources, but assuredly we shall get too its social discords and moral plagues and cruel problems, and I do not see how we shall avoid becoming the slaves of the economic aim in life and losing the spiritual principle of our culture.

Yet, it is not an Indian policy–making bureaucrat or politician or economist of today from whom we can get an echo of the Aurobindovian prophecy. For example, in 1995 the redoubtable jurist N.A. Palkhivala waxed eloquent about the trinity of privatization–liberalization–globalization.[43] It is a Professor of European Thought at the LSE who tells us now:[44] 'The world into which we are moving is an increasingly dangerous place, but at least we can begin to think about its risks realistically without the disabling illusions of the era of globalization'. How true is the Indian proverb: the yogi does not get alms in his own village! Sri Aurobindo was a yogi – wasn't he ?

However, from among the few honourable exceptions at a level whose opinions count for something, we may refer to the eminent scientist M.S. Swaminathan. Unlike Palkhivala, he has not minced words or uttered platitudes:[45]

> Right now, as a result of the WTO and indiscriminate import of agricultural commodities, our farmers are struggling to defend their home market, forget capturing new markets ... Therefore the WTO is not a level playing field at all, especially in agriculture.

Huntington again, like Gray, extends the perspective on economic globalization over a wider cultural canvas.[46] While we have some honest here-and-now realism of a grassroots character in what Swaminathan, Huntington and Gray say today about globalization, in Aurobindo we had a preview of the entire scope of the problem from the height of yogic holism – decades ago.

Notes

1 *Guardian Weekly*, 8–14 March 2001, p. 12.
2 Henderson, H. (1999) *Beyond Globalization*. Connecticut: Kumarian Press, p. 2.
3 *The Economist*, 23 September 2000, pp. 17–18.
4 For example, *Globalization: The Perspectives and Experiences of Religious Traditions of Asia-Pacific*, (eds), J.A. Camilleri and C. Muzaffar. Malaysia: International Movement for a Just World, 1998.
5 Rodrick, D. (1997) *Has Globalization Gone Too Far?* Washington, DC: Institute for Economics.
6 *Forbes ASAP*, November 1998, vol. 162, no. 12, p. 69.
7 Sri Aurobindo, *The Message of the Gita*. Pondicherry: Sri Aurobindo Ashram, 1977, p. 206.
8 *Ibid.*, p. 228.
9 *Ibid.*
10 *World Happiness Survey*, NRIOL Cam Report, 14 December 1999.
11 *Ibid.*
12 SIDH, Society for Integral Development of Himalayas, *A Matter of Quality*. Mussoorie: Sansodhan Research Paper, 1999, p. 8.
13 *Ibid.*
14 *Ibid.*, p. 18.
15 *Ibid.*, p. 35.
16 Shurner Smith, P. (2000) *India – Globalization and Culture Change*. London: Arnold, p. 32.
17 *Ibid.*, p. 33.
18 *Ibid.*, p. 35.
19 Rugman, A. (2000) *The End of Globalization*. London: Random House, p. 19.
20 Coates, B. 'Big Business At Your Service', *Guardian Weekly*, 15–21 March, p. 28.
21 *Crossing Boundaries*, Spring–Summer, 1999, p. 13.
22 Rugman, A *op. cit.*, p. 77.
23 *Ibid.*, p. 78.
24 *Ibid.*, p. 81.
25 *Ibid.*, p. 83.
26 Coomaraswamy, A. (1989) *What is Civilization?* New Delhi; Oxford University Press, p. 7.
27 *Ibid.*, p. 83.
28 Kimber, D. (1999) 'Ethics in International Technology Transfer', in S.K. Chakraborty and S.R. Chatterjee (eds), *Applied Ethics in Management*. Hamburg: Springer Verlag, pp. 12–13.
29 Gerle, E. (2000) 'Contemporary Globalization and its Ethical Challenges', *Ecumenical Review*, vol. 52 (2), April, pp. 158–71.
30 Rugman, A *op. cit.*, p. 9, pp. 217–18.
31 *Questioning Globalization*, the Christian Century Foundation, October 1998.
32 Hoogvelt, A. (1997) *Globalization and the Post-colonial World*. London: Macmillan, p. 14.
33 *Ibid.*, pp. 115–25.
34 Quoted in D.C. Korten, *The Post-Corporate World*. San Francisco: Berrett Koehler & Kumarian, p. 119.

35 Rabindranath Tagore, *Lectures and Addresses*. New Delhi: Macmillan, 1988, p. 72.
36 *Ibid.*, p. 66.
37 Beer, S. (1994) *How Many Grapes Went into the Wine*, reprinted in a collection edited by R. Harnden and A. Leonard. London: John Wiley, p. 321.
38 Sri Aurobindo, *The Human Cycle*. Pondicherry: Sri Aurobindo Ashram, 1972, p. 571.
39 *Ibid.*, pp. 408–9.
40 *Ibid.*, p. 413.
41 *Ibid.*
42 Sri Aurobindo, *The Foundations of Indian Culture*. Pondicherry: Sri Aurobindo Ashram, 1975, p. 388.
43 Palkhivala, N.A. (1995) *Making Indian Industry Globally Competitive*. Bombay: Forum of Free Enterprise, pp. 7–12.
44 Gray, J. (2001) 'Goodbye to Globalization', *Guardian Weekly*, 8–14 March 2001, p. 13.
45 Swaminathan, M.S. (2000) 'A Better Harvest', *Management Review*, December 2000, pp. 98–9.
46 Huntington, S.P. (1997) *The Clash of Civilizations*. New Delhi: Penguin Books, p. 57.

15
Freedom and Core Values for Indian Development

Bhanoji Rao and M.V. Lakshmi

I give you what you want so that you may learn to want what I have to give – Prema (Love) and Liberation itself.

(Bhagawan Sri Sathya Sai Baba,
from the web site of the
Sri Sathya Sai Institute of Higher Medical Sciences)

Constituents of development

The conceptualization of development has changed substantially in the past four to five decades. From a pure increase in the volume of goods and services per head, we have come a long way to include per capita income, literacy and life expectancy in the human development idea popularized by the United Nations Development Programme in its annual *Human Development Reports*. To a large extent, the human development idea and the human development index based on that have received appreciation and approval from most development economists. This is because of the simple recognition that for an individual to function and lead a decent life, she/he needs a certain minimum purchasing power, good health and knowledge. These three in fact are the building blocks of the individual's capacity to lead a decent life. Together they indicate human capability.

Leading a life of one's choice is often not possible without individual freedom and the institutions safeguarding that freedom such as a free press, periodic elections for leadership positions at all levels of government and a system of governance that allows for participation in decision-making. It is thus well-recognized in the relatively more recent literature that both human capability and freedom are integral to development.

Freedom enables a person to exercise choices (including consumption choices). It allows her/him to voice opinions, participate in political and other debates, and hold political office, regardless of the person's attributes such as religion, economic status and so on. Freedom of the people enshrined in a democratic state assures sovereignty of the people, accountability of (elected) leaders, rights of the individual and the rule of law. Rule of law and its integrity are vital to safeguard the freedom of the individual and dispense justice without fear or favour.

As Sen (1999) points out, development is freedom and freedom is development. It is an end as well as the means. As it develops, it helps development. It helps development not only via the freedom to pursue one's education and occupation but also through creativity in all fields.

India in development transition

It may be apt to describe the goings on in India as a manifestation of 'a nation of contradictions, held together by invisible strings'. A major contradiction is the persistence of a rather high level of absolute poverty – in the neighbourhood of some 20 to 30 per cent of the people – in a country that boasts of a large pool of technical and scientific manpower and a vibrant democratic polity.[1] Poverty and other social maladies such as gender inequality, inadequate housing and sanitation prevail in spite of much progress achieved in areas such as building a modern industrial economy.

The fact remains that India's achievements are relatively meagre in comparison to an equally populous China, as the data in Table 15.1 indicate. China has a per capita income nearly twice that of India; its exports are around five times that of India; it has relatively less poverty; and its people are relatively better educated and more healthy. Both countries are blessed with vast human resources and considerable natural resources; yet one has taken off while the other is still struggling. China is increasingly considered to be the economic powerhouse that will call the shots in the present century. Every major multinational corporation has a presence in China or is going to have one, most notably to take advantage of the country's vast total buying power (an economy of 4 trillion dollars as against the 2 trillion of India) and the trading opportunities that will go up further upon the country's recent accession to WTO.

One could raise the rhetorical question: of what use is all the development in China in terms of purchasing power, literacy and health if the Chinese are devoid of even minimal freedom of expression? The lack of freedom in China is amply illustrated from the following short

Table 15.1 Selected development indicators for China and India

Indicator		China	India
Population in millions, 2000		1250	998
GNP measured at PPP, 1999: total in billions of dollars		4112	2144
• per capita in dollars		3291	2149
Exports in billions of dollars	(1990)	57	23
	(1998)	207	47
Average annual growth rate, 1980–98 (%)			
• gross domestic product		10.4	5.9
• per capita private consumption		7.2	2.7
Population below $1 a day purchasing power, 1997 or 1998		18.5	44.2
Under–5 mortality rate per thousand children			
	(1980)	65	177
	(1998)	36	83
Adult illiteracy rate (% of people 15 and above), 1998			
	(males)	9	33
	(females)	25	57

Source: *World Development Report*, 2000/2001, Washington, DC: The World Bank, 2000.

extract from Amnesty International's *Annual Report on China for the Year 2000*:

> 1999 saw the most serious and wide-ranging crackdown on peaceful dissent in China for a decade. Thousands of people were arbitrarily detained for peacefully exercising their rights to freedom of expression, association or religion. Some were sentenced to long prison terms under draconian national security legislation and after unfair trials; others were assigned without trial to up to three years' detention in 're-education through labour' camps. Torture and ill treatment of prisoners were widespread. Thousands of people were sentenced to death and many executed. In the autonomous regions of Tibet and Xinjiang those suspected of nationalist activities or sympathies continued to be the targets of particularly harsh repression.

In contrast to the people of China, Indians enjoy much greater political freedom; yet, there are human-rights violations in India too. For instance, Amnesty International's *Annual Report on India for the Year 2000* has this to say:

> Human rights violations occurred throughout India against a backdrop of political instability. The socially and economically weaker sections

> of society continued to be particularly vulnerable to human rights abuses. Attacks, often with the apparent connivance of police and local authorities, on *dalit* communities (disadvantaged groups determined by caste hierarchies) and tribal people were commonplace. Women continued to be particularly vulnerable to abuse in these contexts. Access to justice for these victims of human rights abuses remained problematic and those engaged in protecting the rights of the most vulnerable groups also came under increasing pressure, often themselves becoming the victims of abuses.

It is clear that violence continues in the relatively economically and socially advanced but not quite democratic China as well as democratic India. There are bound to be major differences between India and China in regard to the reach, intensity and effect of the violence perpetrated; however, for the purpose of this essay, it is enough to note that there is evidence of widespread violence in both countries, regardless of who or which agency is the perpetrator of violence.

China and India share not only cases of human-rights violation but also a fairly close corruption perception index (CPI). The index devised by Transparency International ranges from 0 (most corrupt) to 10 (least corrupt). Finland is perceived to be least corrupt (score 9.9), while Bangladesh is perceived to be the most corrupt nation (score 0.4 in 2001). The scores for China and India respectively are 3.5 and 2.7. It is surprising that China, well-known for disciplined people, nationalistic spirit and socialist credentials should have such a poor score. It is ironic that India, with its multiple political parties, adult franchise, parliamentary democracy, activist media, and so on had obtained a score even less than that of China.

China also figured in the 1999 Bribe Payers Index (BPI), compiled by Transparency International for 19 leading exporting economies of the world. The index is based on surveys undertaken in 14 emerging market economies.[2] The data reveal that companies from some of the major exporting economies such as Malaysia, Taiwan, Korea and China are among the top bribe payers. China with a BPI of 3.1 occupied the top spot among all the 19 economies, while scores above 5 are obtained mostly by the industrialized economies.

Despite the fact that China has achieved commendable economic and social progress, the country could not get rid of the relatively high level of corruption. One might in fact argue that economic growth would create expanded opportunities for corruption, as instances from China as well as several of the East Asian case studies in Backman (1999) indicate.

Volumes could be written on corruption if one could obtain all the corruption episodes in the world.[3] Long life, education and relatively high purchasing power, thus, may not lead to reduced corruption.

If corruption continues even in the context of rapid economic growth and high levels of human development, where then does the blame lie? In the wake of the recent (1997–98) East Asian crisis, many have argued that democratic forms of governance will go a long way in combating corruption.[4]

Does democracy promote accountability and minimize corruption? The question gains added significance in the light of the fact that vast parts of the world in the twentieth century have seen the rise of freedom and democracy. Also, as reputed scholars like Fukuyama (1999) assert, market-driven economies and freedom-driven politics seem to be the eventual destiny of the great many nations in the years to come. This projection is also supported by the fact that at the present time, at the dawn of a new century, there can be none who would openly make a case for dictatorial and undemocratic regimes. If democracy is the destiny of the world, and if it can guarantee accountability and little or no corruption, then the Chinese could hope to be more and more free from corruption as they move towards greater political and individual freedom.

The argument that freedom in general and democratic governance in particular could reduce corruption does not seem to hold when one considers the case of India. On the contrary, the corruption-perception index shows more (perceived) corruption in India than China. In fact it is no exaggeration to say that there are bound to be some people in India who opine that the freewheeling democracy and lack of discipline have contributed to corruption.[5]

If both authoritarian China and democratic India have the menace of corruption, the root cause must rest not in the political system, but elsewhere – in the lack of routine manifestation of truth.

An important principle of Confucian thought (Zhang, 1999) that has served China (and other nations of East Asia) is the recognition given to knowledge and virtue rather than caste, colour or creed. The twin accomplishments of strict adherence to both knowledge and virtue should produce material prosperity as well as incorruptibility as well as peace and order. If corruption is still rampant in China, the reason must be seen in lack of virtue rather than lack of knowledge. In the absence of virtue, there may even be some misuse and abuse of knowledge.

Lack of adherence to truth in normal life should explain why China, the cradle of Confucian ethics, and India, the land of some of the greatest religious and spiritual leaders and divine beings, both share the curse of corruption. To expect people to lead lives of non-violence and truth is not utopian. We have seen Mahatma Gandhi and Mother Theresa living lives manifesting the highest human values.

Redefining and reorienting development

The principal engine that nurtures economic development is the work and creativity of individuals. What induces them to strive and invent is a climate of freedom that provides avenues for shaping their own destiny. The alternative situation, that of submission or rebellion, the manifestation of both widely prevalent in India today, is counter-productive and hampers development. Be it business or government, the climate of freedom and the culture of the practice of truth and non-violence set the overall framework for progress via not only hard work but also innovation, the two engines of economic and social progress.

The values prevailing at the crucial moments of decision making in business and government establishments have a crucial role leading to economic development. For instance, devoid of the core values, decisions taken could favour some undeserving people, groups or regions. While such processes could be taken as developmental achievements in the short run, they could in fact be promoters of social tensions and instabilities. In view of this, the practice of core values has to become a part of each decision-maker's personality.

It is thus no longer enough to think in terms of the hitherto discussed constituents of development. In addition to health, education, purchasing power and freedom, which are widely acknowledged as the constituents of development in contemporary literature, it is necessary to add two core human values – non-violence and truth – explicitly as additional constituents.

Between the two core values – non-violence and truth – non-violence takes a higher place in the event of a tie between the two in practical life. While the full manifestation of the two values, which may be called human perfection, may remain an ideal, with reference to the vast majority of diverse societies, what is sought in rising levels of development is their increasing manifestation.

With thousands of commodities and services offered in the marketplace and numerous persuasive advertisements, it is natural to expect that the producers and sellers adhere to truth in order that the markets function as expected. Free markets are important social organisations. They cannot function properly and efficiently in the absence of truth. Similarly, the functioning of free societies requires that people come together on many occasions in smaller or larger groups to deliberate, discuss and decide on issues of common concern. It is implicitly assumed that the members of the group remain civilized and conduct business honestly and truthfully with least violence.

There is hardly any meaning in canvassing for free debate when there is fear all around and suspicion of possible eruption of violence. Also, there is hardly any value for freedom of expression if utter lies are propagated as facts. In the absence of truth and non-violence, freedom may mean nothing but fear and worry about security and safety for one-self and one's near and dear.

In line with the revised conceptualization of development, a society moving towards higher levels of development will have increasing manifestations of non-violence in practical life. In a similar way, as a society moves up on the development scale, the population must be seen to be adhering to truth increasingly. The relatively more developed society will thus have, for instance, not only less crime but also more people admitting the crimes they have committed.

In summary, therefore, *development is the process of achieving the trinity of human capability, freedom and movement towards human perfection* (Rao, 2001). The movement towards human perfection is the outcome of the person increasingly manifesting non-violence and truth in daily life. Increasing manifestation of truth and non-violence should naturally promote altruistic behaviour in increasing numbers of men and women. That in turn would lead to their producing as per their ability and releasing for social purposes the excess of earnings over a comfortable threshold income. This will eventually reduce income inequality.

Similarly, the increasing manifestation of truth and non-violence promotes the awareness of the transient nature of one's possessions, leading to a growing transfer of private property to social purposes and a reduction in wealth inequality, without violence and murder, characteristic of the erstwhile communist revolutions. A relatively more egalitarian income and wealth distribution is thus the likely byproduct from the interaction of non-violence and truth with market forces. They in turn promote peace, and social cohesion and stability.

Development, which comprises the trinity of human capability, freedom and the movement towards human perfection, allows individuals and groups to lead *happy and contented lives and not some happy and contented moments*. Contentment (along with peace of mind) is the final outcome sought from long and healthy life, education, purchasing power and freedom. However, for achieving the contented (and peaceful) state of mind, one also requires knowledge and practice of non-violence and truth, not only by the person concerned, but also by others in the society. In their absence, the person will be devoid of contentment (and peace), despite high levels of education, health, purchasing power and freedom.

Just as material progress (education, health and purchasing power) cannot automatically lead to freedom, and just as freedom does not necessarily provide the wherewithal for decent living, neither material progress nor freedom can ensure that the human being adheres to non-violence and truth. It is, therefore, important to bring the core human values into the mainstream of the development discourse.

Finally, the issue of the purpose of life must be raised and related to the concept of development. The very first Confucian principle (see Zhang, 1999 and Rao, 2001) enshrines what may be called the supreme objective of human life: finding or experiencing the universal truth or the 'Way' for one-self. It would seem to be the same as the concept of abiding in the Self, enunciated by numerous Indian texts and spiritual practitioners. If it is said that the objective of life is self-realization, then one must be provided with the wherewithal for achieving that objective. Knowledge, good health, minimum purchasing power and freedom are only one part of that armoury; truth and non-violence are the other complementary part.

It would appear that the aim of a highly evolved Confucian society is not just material progress; such progress is the means and the goal is self-realization. The contact with the true Self (the 'I' that observes and the witness of all) is not possible via material progress alone. Material progress, however, can be the basis for the next stage – the stage of the inquiry into the true Self. It is proclaimed by the spiritual masters that the vision of the Self and the springing forth of love and compassion happen naturally, automatically and simultaneously. It is the end of finding the 'Way'.

The fact of the matter is that neither India nor China has produced a mass movement towards human perfection. That is yet to come via the conscious and routine adherence to non-violence and truth.

In the absence of adherence to truth and non-violence, especially in the leadership and bureaucracy, Indian development will stall despite liberalization of the economic system and in spite of the system of democratic governance. It is important to note that systems are designed and operated by people. The extent to which people designing and operating the systems adhere to core human values, sets a limit to the efficient functioning of the systems they design and operate.

Notes

1 In February 2001, based on a 30-day consumption methodology, the National Sample Survey Organization (NSSO) data revealed that the poverty level stood at 26.10 per cent of the population as against 36 per cent estimated by the 1993–94 NSSO survey. On the one hand the current level was still higher than the target of 16 per cent set for the Ninth Plan, while on the other, the controversy on trends and levels of poverty continues. For a recent exposition on the controversy, see Lal, Mohan and Natarajan (2001).

2 The surveys involved detailed questions to senior executives at major companies, chartered accountancies, chambers of commerce, commercial banks and law firms. BPI ranks the bribe-payers in terms of the degree to which their corporations are perceived to be paying bribes abroad. The questions concerned the propensity to bribe senior public officials by corporations. The 14 countries are: India, Indonesia, Philippines, South Korea and Thailand (Asia/Pacific); Argentina, Brazil and Colombia (Latin America); Hungary, Poland and Russia (Europe); and Morocco, Nigeria and South Africa (Africa). These economies account for more than 60 per cent of the total imports of all emerging economies put together.

3 Even Taiwan, which gained the reputation of a high growth and low inequality economy, is not free from corruption. For instance, media reports have publicized a financial scandal that erupted in Taiwan in early 1999 involving a former speaker of the legislature who was also the chairman of a major bank. It is alleged that money and political power go together in Taiwan despite its accomplishments in human capability.

4 See, for instance, Backman (1999) and Ramos (1999).

5 It can be also argued that a benevolent dictator, good at enforcing contracts, safeguarding property rights, and facilitating free markets, could provide the necessary conditions for rapid economic and social development. For the sake of popular support she/he could usher in several egalitarian measures and projects such as assured housing for all, and good sanitation and water supply for the entire population.

References

Backman, M. (1999) *Asian Eclipse: Exposing the Dark Side of Business of Asia*. Singapore: John Wiley & Sons.

Fukuyama, F. (1999) 'Asian Values in the Wake of the Asian Crisis', paper presented at the *International Conference on Democracy, Market Economy and Development*, Seoul, 26–27 February 1999.

Lal, D., Mohan, R. and Natarajan, I. (2001) 'Economic Reforms and Poverty Alleviation: A Tale of Two Surveys', *Economic and Political Weekly*, vol. 36(12), 24 March, pp. 1017–28.

Ramos, F.V. (1999) 'Democratization and Economic Development – An East Asian Perspective'. Paper presented at the *International Conference on Democracy, Market Economy and Development*, Seoul, 26–27 February 1999.

Rao, B. (2001) *East Asian Economies: The Miracle, a Crisis and the Future*. Singapore: McGraw-Hill.

Sen, A. (1999) *Development as Freedom*. Oxford: Oxford University Press.

Zhang, W.-B. (1999) *Confucianism and Modernization: Industrialization and Democratization of the Confucian Regions*. London: Palgrave Macmillan.

Index